Smart Management

Smart Management

How Simple Heuristics Help Leaders Make Good Decisions
in an Uncertain World

Jochen Reb, Shenghua Luan, and Gerd Gigerenzer

The MIT Press
Cambridge, Massachusetts
London, England

The MIT Press would like to thank the anonymous peer reviewers who provided comments on drafts of this book. The generous work of academic experts is essential for establishing the authority and quality of our publications. We acknowledge with gratitude the contributions of these otherwise uncredited readers.

This book was set in Stone Serif and Stone Sans by Westchester Publishing Services. Printed and bound in the United States of America.

Library of Congress Cataloging-in-Publication Data

Names: Reb, Jochen, 1973– author. | Luan, Shenghua, author. | Gigerenzer, Gerd, author.
Title: Smart management : how simple heuristics help leaders make good decisions in an uncertain world / Jochen Reb, Shenghua Luan, Gerd Gigerenzer.
Description: Cambridge, Massachusetts : The MIT Press, [2024] | Includes bibliographical references and index.
Identifiers: LCCN 2023027153 (print) | LCCN 2023027154 (ebook) | ISBN 9780262548014 (paperback) | ISBN 9780262378567 (epub) | ISBN 9780262378574 (pdf)
Subjects: LCSH: Decision making. | Management. | Leadership.
Classification: LCC HD30.23 .R42 2024 (print) | LCC HD30.23 (ebook) | DDC 658.4/03—dc23/eng/20230925
LC record available at https://lccn.loc.gov/2023027153
LC ebook record available at https://lccn.loc.gov/2023027154

10 9 8 7 6 5 4 3 2

For Franziska, Waltraud, and Werner
For Peiqiu, Huizhen, and Tian
For Raine, Thalia, Kyle, and Athena

Contents

Acknowledgments   ix

Part I

1 What You (Likely) Won't Learn in Business School   3
2 Why Heuristics?   15
3 The Adaptive Toolbox   33

Part II

4 Hiring and Firing   55
5 Strategy   73
6 Innovation   89
7 Negotiating in the Real World   103
8 Building Better Teams and Communities   121
9 Leaders' Adaptive Toolbox   133

Part III

10 The Power of Intuition   151
11 Creating Smart Decision-Making Cultures   161
12 Artificial Intelligence and Psychological Intelligence   177
13 What You Should Learn in Business School   195

Glossary   211
Notes   217
References   231
Index   251

Acknowledgments

Smart Management: How Heuristics Help Leaders Make Good Decisions in an Uncertain World was conceived in the large world, under uncertainty. The COVID-19 pandemic presented a challenge: with the three of us based in three different countries and time zones, we could not work together in person as much as we would have liked to but had to settle for a lot of virtual meetings. That said, as is common for a world of uncertainty, the situation also presented an opportunity: more time at home and in quarantine meant more time for thinking and writing. We tried to make the best of difficult circumstances and are glad that we could work together on a project that all three of us feel very passionate about: how managers, leaders, and organizations can make better decisions in a large world with the help of smart heuristics.

This book developed out of a review paper we coauthored for an academic audience, *Smart Heuristics for Individuals, Teams, and Organizations* (Gigerenzer, Reb, & Luan, 2022). Believing that the ideas in that review were relevant not only to other academics but also to management practitioners and educators, we decided to invest the time to write this book. The conversation unfolded something like this: In our first meeting after completing the page proofs of the review article, Gerd asked "So, what next?" Neither Shenghua nor Jochen had given this question much thought. Gerd, in his wisdom, used the ensuing pause to drop what turned out to be a consequential proposal: "What about turning this into a book?" Hmm, why not?! It sounded like a good idea at the time—promising enjoyable intellectual collaboration, bridging the science–practice gap often lamented in management, and, we hoped, making a positive impact by providing useful ideas for making smart decisions. It still sounds like a good idea now, despite (or perhaps because of) the countless hours of writing, online meetings, and intensive on-site writing retreats that followed.

We are immensely grateful for the support that we have received throughout this project (and the earlier review paper) from individuals and institutions around the world. For the stimulating and collegial environments, we thank our institutions: the Lee Kong Chian School of Business, Singapore Management University; the Institute of Psychology, Chinese Academy of Sciences; and the Max Planck Institute for Human Development.

For helpful feedback on chapters of the book (and/or the earlier review article), we are deeply grateful to Florian Artinger, Devasheesh Bhave, Kathleen Eisenhardt, Ulrich Hoffrage, Konstantinos Katsikopoulos, Filip Lievens, Theodore Masters-Waage, Thomas Menkhoff, Eva Peters, Michael Schaerer, and participants of the 2022 ABC research retreat. For their careful editing of the entire book, we express our heartfelt thanks to Rona Unrau and Anita Todd. For creating the illustration on leader mismatch in chapter 4, we thank Kikuko Reb and Kai Reb. We are thankful to Markus Buckmann and Nikita Kozodoi for sharing their data, and to Ying Li, Jun Liu, and Yuhui Wang for help in making some of the figures.

Last but not least, we would like to thank our families for their emotional, intellectual, and logistic support. To Lorraine Daston, thank you for graciously hosting at least one of us (Reb) during two intensive writing retreats. To Athena, thank you for providing entertainment during our writing breaks. To Kikuko, Kai, and Ken Reb, thank you for patiently accepting your husband and father spending additional time away from family to work on this book. To Tian Liu and Julian Luan, thank you for all the conversations with your husband and father on writing, heuristics, and decision making that provided some fun ideas for this book.

Part I

1 What You (Likely) Won't Learn in Business School

Remember the COVID-19 pandemic, the subsequent disruption of global supply chains, and the resulting product shortages? Before the pandemic, many institutions clung to the *illusion of certainty*—the belief that the world is more certain than it actually is. After decades of holding fast to the assumption that all risks can be foreseen, managed quantitatively, and controlled, an appreciation of the reality of uncertainty is reentering corporate and government offices. The distinction between risk and uncertainty lies at the heart of making effective decisions using smart heuristics.

Business schools teach plenty of skills, but they tend to leave out some of the most useful ones when it comes to making decisions. That omission is not accidental. Courses on management, leadership, and finance teach aspiring managers that rational decision making means choosing the alternative with the highest expected utility, which requires foreseeing all the possible consequences of each potential option. Good managers, so the story goes, search for all relevant options, carefully analyze all the possible consequences, weight utilities by probabilities, and calculate which option maximizes the expected utility. Around the world, business schools teach this procedure to legions of students. "More is better" has become an article of faith: more data, more information processing, and more analysis are all assumed to improve decision making.

Long ago, Benjamin Franklin advised a nephew who was considering marriage: If in doubt, list the pros and cons of all your options, weight them, and do the calculation; otherwise, you will never get married.[1] Yet few people actually choose their partners by doing a calculation—and rightly so. Finding a suitable partner involves a high degree of uncertainty, as divorce rates indicate. When it comes to marriage, one cannot foretell *all* the possible consequences, not to mention their precise probabilities. The same holds true in business: it is impossible to foresee all possible consequences of entering a foreign market, acquiring a company, or hiring a new CEO.

In general, the expected utility maximization procedure that business schools teach and Benjamin Franklin advocated is useful in stable, well-defined situations where nothing unexpected ever happens. However, executives operate in an increasingly *volatile, uncertain, complex, and ambiguous* (VUCA) world. Here, the advice to collect all information, consider all options, and anticipate all the possible consequences and their associated probabilities is of little use. It creates an illusion of certainty.

Nonetheless, managers regularly decide whom to hire, when to terminate a project, and whether to acquire another company. To make these decisions, they rely on a set of tools called *heuristics*. Surprisingly, business schools rarely instruct their students in how to use these powerful tools to make intelligent decisions. Instead, if heuristics are mentioned at all, they are portrayed as something to avoid in favor of more complex decision strategies. Popular science books echo this negative view and tend to attribute (in hindsight) all kinds of disasters, from obesity to financial crises, to "heuristics and biases."[2] In this book, we share with you a more positive, realistic, and practical view and provide a systematic introduction to the science and art of heuristic decision making.

A *heuristic* is a simple rule that enables decisions to be made quickly, frugally, and accurately. Heuristics are necessary tools in situations of uncertainty (i.e., when the conditions required for maximizing utility do not exist). The distinction between situations of risk, where maximizing is possible, and situations of uncertainty, where it is not, goes back to the economist Frank Knight.[3] It has since been mentioned in virtually every economics textbook, only to be subsequently ignored. Here, we give uncertainty the attention that it deserves, and consequently, we take heuristics seriously.

We begin by introducing three Nobel Memorial Prize laureates in economics. What did they think about decision making? And how did they actually make their own decisions?

Herbert Simon and Satisficing

In 1978, Herbert A. Simon was awarded the Nobel Memorial Prize in Economic Sciences "for his pioneering research into the decision-making process within economic organizations."[4] The *process* of decision making can determine the rise or fall of an organization. Strikingly, however, this very process is rarely considered relevant in theories of organizations and economics. Instead, economic theory posits that managers behave as if they maximize their expected utility, no matter how they make decisions. Simon rejected the assumption made in classic theory that executives are omniscient profit or utility maximizers, and he highlighted the total lack of evidence that the theory describes how decisions are actually made. Responding to this type of criticism, Milton Friedman famously stated back in 1953 that it is immaterial whether expected utility maximization describes the process of decision making or not; it is simply a tool for predicting behavior, and all that counts is its predictive accuracy.

However, a review of fifty years of research on utility functions—including the utility of income functions, the utility of wealth functions, and the value function in prospect theory—concluded that the power of utility functions "to predict out-of-sample is in the poor-to-nonexistent range."[5] This finding supports Simon's criticism that expected utility theory not only fails to describe how decisions are made but also is too indeterminate and flexible to predict well.

Out-of-sample means that predictions are made beyond the data used to create a model in the first place. In contrast, utility and other complex models are often tested by simply fitting their parameters to already-known choice data. It is sometimes misleadingly said that these models "predict" decisions when in actuality, they optimize the fit to past data. *Optimizing* is a mathematical concept that means determining the maximum or minimum of a curve, such as a utility curve. The distinction between fitting and predicting is crucial. A more complex model with more free parameters can obviously fit past data better. However, often the resulting model overfits and is less able to predict—for instance, when the future is not like the past.[6] Thus, even if one were to accept Friedman's questionable argument that theories need only predict outcomes, not describe the process of behavior, expected utility would still fare poorly: it neither describes nor predicts well, at least not in the uncertain world of business.

Simon proposed instead that decision makers *satisfice* in situations where optimization is impossible. The term *satisficing*, which originated in Northumbria (a region in England on the Scottish border), means "to satisfy." Simon learned about satisficing from direct experience: In the mid-1930s, fresh from a class on economic theory at the University of Chicago, he tried to apply utility maximization to budget decisions in his native Milwaukee's recreation department. To his surprise, he learned that managers did not compare the marginal utility of a proposed expenditure to its marginal costs. Instead, they simply added incremental changes to last year's budget, engaged in habits and bargaining, or voted on the basis of their identification with organizations. Simon concluded that in the real world of business, the framework of utility maximization "was hopeless."[7] This experience led him to a new question: How do human beings reason when the conditions for rationality postulated by the model of neoclassical economics (such as having complete information and operating under risk rather than uncertainty) are not met? He found the answer in heuristic processes, including recognition, satisficing, heuristic search, and aspiration levels. The study of heuristics enables us to both describe the process and predict the outcome of decisions.

Simon lived what he taught about satisficing. He made decisions easily and rapidly, considering only a few options and their key consequences. According to his daughter, Kathleen Simon Frank, he was always willing to live with the outcomes of his decisions rather than constantly mulling over things.[8] He was very down-to-earth, with a frugal lifestyle, wearing the same type of clothes every day. He had just three shirts: one that he wore, one in the wash, and one in his closet. Simon lived a satisficing life, using the time that he saved that way to indulge his love of reading widely in the sciences, listening to his students, and debating ideas.

Harry Markowitz and the 1/N Rule

Twelve years after Simon, in 1990, Harry Markowitz received the Nobel Memorial Prize in Economic Sciences for his theory of portfolio choice. Unlike Simon's satisficing, which described how people *actually* make decisions, Markowitz's theory was normative; that is, it prescribed how investors *should* allocate wealth in assets that differ in return (mean) and risk (variance). Markowitz was honored for developing a mathematical formula, the mean–variance portfolio, that aims at maximizing profit. Business schools all around the world continue to teach his formula and its many variants. The method requires an exhaustive analysis of financial data to predict future returns, variances, and covariances. For a large number of assets, this may require estimating thousands or even millions of numbers.

When Markowitz made his own investments for the time after his retirement, one might assume that he followed his own Nobel Prize–winning formula. In fact, he relied on a heuristic known as the *1/N rule*: invest your money equally across *N* options. If $N = 2$, this means a 50:50 allocation, and so on. In behavioral finance, relying on $1/N$ is dubbed *naive diversification bias* and is attributed to people's cognitive limitations and irrationality. But that clearly does not apply to an economist with the stature of Markowitz. He later explained that his decision to invest equally in stocks and bonds aimed at avoiding future regrets: "You know, if the stock market goes way up and I'm not in it, I'll feel stupid. And if it goes way down and I'm in it, I'll feel stupid. So I went 50-50."[9]

Those who think of $1/N$ as naive overlook an important fact. Markowitz's portfolio choice theory is optimal only in a world where one can foresee all future returns, variances, and covariances of all assets. In an uncertain world, however, estimating thousands or millions of parameters from past data leads to overfitting. The $1/N$ rule, by contrast, has no free parameters, so it cannot overfit. What appears to be a limitation—no free parameters—can turn out to be an advantage in prediction. Accordingly, subsequent studies showed that in stock investments, $1/N$ often outperforms the mean–variance portfolio because it is more robust and does not overfit.[10] Less can be more.

Reinhard Selten: Game Theory and Real-World Problems

Four years after Markowitz, in 1994, Reinhard Selten won the Nobel Memorial Prize in Economic Sciences for his work on game theory. Yet he had two scientific passions: alongside game theory, the psychological study of heuristics and bounded rationality. As a trained mathematician, he made a clear distinction between the two. He regarded his work on game theory as a mathematical exercise in a well-defined world, *not* as how people do or should behave in the real and uncertain world.

Consider Selten's famous chain store paradox:[11] A chain store called Paradise has branches in twenty cities. A competitor, Nirvana, plans to open a similar chain of stores and to decide one by one whether to enter the market in each of these cities. Whenever a local challenger enters the market, Paradise can respond either with aggressive, predatory pricing, which causes both sides to lose money, or with cooperative pricing, which will result in sharing profits fifty-fifty with the challenger. How should Paradise react when the first Nirvana store enters the market—with aggression or cooperation?

Selten proved that the best answer is cooperation. His proof is based on the principle of *backward induction*, in which one logically argues backward from the end to the beginning. When the last of the twenty challengers enters the market, there is no reason for aggression because there is no future competitor to deter, and thus one should cooperate and not sacrifice money. Now consider what to do with the next-to-last challenger. Given that Paradise will be cooperative toward the twentieth challenger, there is no reason to be aggressive toward the nineteenth either because everyone knows that the chain store will cooperate with the final challenger. Thus, Paradise should cooperate with this challenger too. The same argument applies to the eighteenth challenger, and so on, all the way back to the very first. Selten's proof by backward induction implies that in every city, the chain store should respond cooperatively from the first challenger to the last.

After deducing the result, Selten found his proof intuitively unconvincing and indicated that in the real world, he would instead follow his gut feeling to be aggressive to deter others from entering the market:

> I would be very surprised if [aggression] failed to work. From my discussion with friends and colleagues, I get the impression that most people share this inclination. In fact, up to now I met nobody who said that he would behave according to [backward] induction theory. My experience suggests that mathematically trained persons recognize the logical validity of the induction argument, but they refuse to accept it as a guide to practical behavior.[12]

Selten's disclosure might lead some to suspect that he was a person whose impulses overwhelmed his thinking. But the real explanation of his dismissal of the logical conclusion lies in his discerning between well-defined situations with complete information, such as in the chain store problem, and the ill-defined reality of business competition, where backward induction is no longer a safe guide. The clash between Selten's logic and his intuition is a consequence of this important conceptual distinction.

Up to the present day, however, most business schools continue to teach that logical arguments are the benchmark for effective business decisions and relying on heuristics and expert intuition will lead to failure. Consider a panel discussion held at the prestigious biannual OWL Science-Meets-Entrepreneurs Day at the Faculty of Business Administration and Economics at Bielefeld University, where Selten and one of us (Gigerenzer) spoke with two successful local entrepreneurs. The assembled audience expected to witness a heated debate between scientists and entrepreneurs, but when Selten and Gigerenzer spoke about the benefits of heuristics and the importance of intuition for innovation, the two entrepreneurs wholeheartedly agreed. Exploiting simple rules and gut feelings was the way that they built up their companies and made their fortunes, although they had learned little about either in business school. In the end, Selten and Gigerenzer appealed to the professors in the audience to take uncertainty and heuristics seriously and to start teaching their students how people make profitable decisions in the real world. At first, the audience was speechless in shock, and then they nearly hit the ceiling. But the two entrepreneurs rose to the defense, insisting that little of what they had learned in business school was of practical use for their business careers. Visibly amused by the debate was the president of the university, a computer scientist by training, who knew that heuristics are the bread and butter of programming.

Uncertainty Is Not Risk

These three Nobel Memorial Prize laureates represent three approaches to decision making. Simon proposed theories of heuristic decision making and used them to make his own decisions. Markowitz proposed theories of optimal portfolio allocation but relied on a heuristic for his own retirement investing. Selten developed both theories of optimal behavior in well-defined games and theories of heuristics in an uncertain world and relied on heuristics and gut feelings for his own decisions. While all three laureates relied on heuristics for their personal decisions, the key difference among their theories is whether these also dealt with heuristic decisions or just with optimizing alone. Optimizing is possible only in a well-defined, stable world with known probabilities—what Knight called "risk." It is meaningless in situations of uncertainty.

Simon, like Knight, distinguished between situations of risk and uncertainty and, with his empirically oriented mind, he was curious about how people make good decisions under uncertainty. As mentioned previously, he learned about the limits of optimization approaches when working on budget decisions, but also when conducting research on artificial intelligence (AI), of which he was one of the founders. To date, he is the only recipient of both the Nobel Memorial Prize in Economic Sciences and the Turing Award, which has been dubbed "the Nobel Prize in Computing." Most interesting problems in computer science are intractable (i.e., no optimal solution can be found), as is the case in the games of chess and Go. Early on, Simon realized that logical solutions, such as backward induction or expected utility maximization, do not work when a problem is intractable. By nevertheless insisting on the mathematics of maximization, as most theorizing in economics does, one is forced to exclude all intractable problems, along with all situations of uncertainty. In this way, game theory ends up excluding virtually all challenging games that people like to play, and expected utility theory becomes inapplicable to real-world business decisions. When performing an exhaustive search is impossible, instead of letting the optimizing framework dictate which problems to study and which not, Simon's approach was heuristic: to study complex games such as chess and investigate how successful players decide upon their next move.

Markowitz's theory assumes such a stable world of risk. The idea is to harness huge amounts of data and estimate future returns, including their variances and covariances. Modern finance originated from his and Robert C. Merton's similar approaches to portfolio allocation. It treats finance as if it were a lottery, not a situation of uncertainty. Merton, another recipient of the Nobel Memorial Prize in Economic Sciences, applied this framework while on the senior management team of the hedge fund Long-Term Capital Management. It did not go well. The fund lost billions in the aftermath of the unexpected Russian financial crisis and had to be bailed out by the Federal Reserve.[13] Optimized portfolios are fragile in an uncertain world: the analysis of past correlations provides a guide to future asset returns only so long as the future is like the past.

Finally, the beauty of Selten's approach is that he studied both situations of risk, as in game theory, and situations of uncertainty and intractability. As the chain store paradox illustrates, he rejected the idea that logical arguments can prescribe how we should act in the real world of business, where neither side has complete information or is obliged to follow the rules of the chain store game. Selten loved game theory because it was mathematically challenging (he was a mathematician, after all), but he did not confuse it with a theory about how we behave, or even how we should behave outside closed worlds. He considered it a mistake to think of expected utility theory as the only rational theory. In fact, the motto of his book *Bounded Rationality* (written with Gigerenzer) was "to study how people make decisions without probabilities and utilities."[14] While many economic theorists are uneasy with the exaggerated assumptions of rational choice theory, they continue to apply them because they do not see a clear alternative. But both Selten and Simon showed that there is an alternative: the study of heuristic decision making.

The Science and Art of Heuristic Decision Making

In this book, we demonstrate how business executives can make good decisions in a VUCA world with the help of smart heuristics. We do so by drawing on the research on *fast-and-frugal* heuristics inspired by the work of Simon and Selten, as well as on observations of how professionals such as Markowitz actually make decisions. We provide real-world examples and offer practical advice for how leaders and organizations can develop their own *adaptive toolbox*, or repertoire, of heuristics to make effective decisions.

In a VUCA world, complex analytic methods quickly reach their limits or become entirely inapplicable. *Less is often more*, and complexity is better tackled with simple strategies. In such environments, simple rules that search for and use little information often lead to better decisions by being not only faster, but also more accurate, transparent, and easier to communicate, teach, and learn. Although practitioners use heuristics on a daily basis and practitioner books in business extol the virtues of intuition and rules of thumb, they do not always understand why and under what conditions heuristics work. This book aims to change that by providing a theory-informed and research-based, yet practical discussion of how business leaders can use heuristics to make good decisions under uncertainty.

For Simon, intelligence was the product of both the inner cognitive system and the outer environment. To succeed, the inner system must be "smart" (i.e., able to exploit features of the environment with its limited capacities); heuristics are embodiments of this general adaptive strategy. This positive view of heuristics, however, diminished substantially after the 1970s, when heuristics became associated with systematic biases in judgments and decisions and were deemed inferior to expected utility models. Although this assumption is generally true in situations of risk, where all probabilities and consequences are known with certainty, it does not hold in situations of uncertainty and complexity, where optimization loses its meaning and where being robust and adaptive is of great importance. Launched in the

1990s, the fast-and-frugal heuristics research program, which we discuss in more depth in the following chapters, has revived and extended Simon's view on heuristics. A multitude of studies have shown that simple heuristics are often superior to complex models.[15]

A Very Short Preview

Building on the fast-and-frugal heuristics program, this book demonstrates the efficacy of heuristic decision making in a twofold approach. First, it describes the *adaptive toolbox*, which leaders, managers, and professionals can use to make decisions. Second, and more important, it introduces the concept of *ecological rationality*, which prescribes the environmental conditions under which specific heuristics work well. Like any strategy, a heuristic cannot work well in all situations, which makes it important to understand, in a principled way, when it will be effective and when not.

In this book, we use the term smart heuristics *as shorthand to refer to heuristics applied in situations where they are ecologically rational. Applied in the wrong context, heuristics can be "not smart," leading to ineffective decisions. Intelligent decision making requires choosing appropriate heuristics for the task at hand.*

Part I of the book provides an introduction to ecological rationality and the adaptive toolbox. Part II describes the adaptive toolbox in areas such as leadership, business strategy, negotiations, and teamwork. Part III covers several cross-cutting themes such as AI and heuristics, the role of intuition, and organizational decision-making cultures.

Using Heuristics—and Feeling Good about Doing So

Executives routinely use heuristics, and yet the misplaced association of heuristics with errors makes them mostly hesitant to admit to it. This hesitancy is generally weaker in family and entrepreneurial businesses, where intuition is more acceptable, and stronger in large corporations and public administrations, where the ideology of optimization dominates. As a result, instead of standing up to their heuristic decisions, executives routinely attempt to hide the actual heuristic decision-making process by creating the appearance that the decision was reached following an exhaustive, quantitative analysis.

Consider a typical case: An executive makes a decision based on a gut feeling, as no clear favorite emerges after considerable deliberation. Afraid to take responsibility for the intuitive decision, the executive instead hires an expensive consulting firm for the purpose of justifying a decision that has already been made with the help of an impressive array of numbers and analytics.

How frequently does that happen in large corporations? When one of us (Gigerenzer) asked the principal of one of the largest consulting firms worldwide how many of the firm's projects involved justifying decisions that had already been made, the response (given on condition of anonymity) was that it was over 50 percent.

Consider how much wasted money, time, and effort could be avoided if organizations took heuristics seriously and studied how and when they work. As a result, they would not have to hide the fact that they regularly used heuristics. Instead, they could feel good about doing so—about making competent decisions in a world of uncertainty. We believe that the time is ripe to revise the image of heuristics in management and business from being biased to being smart.

2 Why Heuristics?

The term *heuristic* is of Greek origin and means "serving to find out or discover." The Gestalt psychologists Max Wertheimer and Karl Duncker used it in this sense, speaking of heuristic methods such as looking around to guide the search for information. Similarly, Albert Einstein included the term in the title of his Nobel Prize–winning 1905 paper on quantum physics to signal that the view that he was presenting was an incomplete yet highly useful route to discovering something closer to the truth.[1] The mathematician George Pólya argued that science requires both analytic and heuristic tools; analysis, for instance, is necessary to check a proof, but heuristics are needed to discover the proof in the first place.[2]

Together with Allen Newell, a student of Pólya's, Herbert Simon introduced heuristic search to make computers more intelligent. The result was the original program of artificial intelligence (AI), which studied the heuristics that experts use and translated these into computer algorithms. Here, the human was the teacher and the computer the pupil. That is why the *I* in *AI* originally referred to human intelligence or, more precisely, human heuristics, acknowledging the fact that heuristics can solve problems that logic and probability cannot. This vision of psychological AI differs from machine-learning systems that rely on brute computing power. Despite their remarkable performance and popularity, these systems have not yet been able to create what could be called human intelligence, and psychological AI is currently being reconsidered as a route to true machine intelligence.[3]

Simon also formulated one of the first algorithmic models of heuristics, known as *satisficing*.[4] Satisficing can lead to good decisions in situations where optimizing is impossible. This view of heuristics as useful tools was turned upside down in the 1970s, however, when researchers began associating heuristics with biases and presented expected utility theory as the universal tool for all decisions.[5] The influence of the heuristics-and-biases program may be one of the reasons why the positive features of heuristics have remained underestimated in management and business.[6] Beginning in the 1990s, the program of fast-and-frugal heuristics took up Simon's unfinished work and extended it by developing algorithmic models of heuristics and introducing the concept of ecological rationality, which refers to the conditions under which heuristics are successful or not.[7] These two important features, algorithmic models and ecological rationality, expand and improve the earlier heuristics-and-biases program: they enable the study of concrete rules that help organizations make better decisions under uncertainty. The two programs should not be seen as antagonistic, but rather as natural steps toward progress.

Heuristics Guide Decision Making under Uncertainty

When are heuristics needed? The key to answering this question is the distinction between *small worlds* and *large worlds*. The term *small world* was coined by Leonard Savage, known as the founder of modern decision theory. Savage made it very clear that the theory of maximizing expected utility applies only to small worlds, and he considered it "ridiculous" to apply it in situations of uncertainty, even those as mundane as planning a picnic.[8] A small world has two features:

1. *Perfect foresight of future states*: The agent knows the exhaustive and mutually exclusive set S of future states of the world.
2. *Perfect foresight of consequences*: The agent knows the exhaustive and mutually exclusive set C of consequences of each of the agent's actions, given a state.

Savage called the pair (S, C) a "small world." A state is "a description of the world, leaving no relevant aspect undescribed."[9] States and consequences must necessarily be described at some limited level of detail, hence the qualifier *small*. A game of roulette constitutes one such small world. All possible future states are known (the numbers 0 to 36 on a European-style roulette wheel), as are all possible actions and their consequences. In roulette, actions include betting on red or black, on even or odd numbers, or on specific numbers or combinations. The complete set of actions need not be mentioned

Table 2.1
The structures of small and large worlds

	Small Worlds		Large Worlds	
Criteria	Risk	Ambiguity	Intractability	Uncertainty
Are all possible actions, future states, and consequences known?	Yes	Yes	Yes	No
Are all probabilities known?	Yes	No	Yes	No
Can optimal action be calculated?	Yes	Yes	No	No

In a small world, all possible actions and future states, along with all possible consequences of each state, are known. If the consequences are not certain but probabilistic and these probabilities are known, the small world is called a *situation of risk*. If the probabilities are not known, it is called a *situation of ambiguity*. Small worlds allow one to calculate the optimal actions; larger worlds do not. *Intractability* refers to a well-defined situation where the optimal action cannot be determined. *Uncertainty* refers to ill-defined situations where not all possible actions, future states, their consequences, and probabilities are known or knowable. Most important problems in computer science are intractable; most important problems in management and decision making involve uncertainty.

separately in (S, C) because actions are defined as combinations between states and consequences. In a small world, everything that can happen is known for certain.

It is of crucial importance to understand that the term *rationality*, as it is used in decision theory and most of economics, is defined for small worlds only. Maximization of expected utility requires such a small world. A small world with known probabilities is referred to as a situation of *risk*, and one without known probabilities as a situation of *ambiguity* (table 2.1). *Intractability* refers to a well-defined situation where the optimal action cannot be calculated. Situations where the state space (S, C) is not fully known or knowable are called *uncertainty*, or sometimes *radical uncertainty*.[10] Under intractability and uncertainty, maximizing expected utility is not an option.

It is sometimes claimed that subjective probabilities could be used here, but that argument fails to distinguish between ambiguity and uncertainty. Subjective probabilities that add up to 1 can be assigned under ambiguity, where the state space is completely known, but not under uncertainty, where it is not. Recall that Savage himself made clear that neither subjective probabilities nor his Bayesian decision theory applies to large, uncertain worlds—not even to ordinary situations such as planning a picnic. In everyday life, small worlds are few and far between. That insight is often ignored. Yet ignoring it does not make expected utility maximization a useful tool for the real world of business.

Thus, a new vision of rationality is needed that is actually useful in large worlds—one that facilitates making decisions under uncertainty, such as for hiring and firing, budget and investment, strategy, and leadership. Uncertainty arises from many unpredictable factors, including human behavior, changes in technology and politics, and personal, financial, and global crises. To make such decisions, smart heuristics are needed. Beyond situations of uncertainty, heuristics are also needed for well-defined situations that are intractable. Examples include scheduling problems such as the traveling salesman problem, where a best sequence of moves exists but no mind or computer can find it. For instance, planning the shortest tour to visit the fifty largest cities in the US is intractable because there are $49! = 49 \times 48 \times 47 \times \ldots \times 3 \times 2 \times 1$ possible tours, which amounts to a number larger than 10^{62} (i.e., a number with sixty-two zeros). To find a good solution in such a vast space, a heuristic search is necessary. An example is the simple *nearest-neighbor heuristic:* Visit the nearest city that has not yet been visited. This heuristic can find excellent solutions to the problem and other similar ones. It is used not only by humans but also by quite a few other species when they forage, including fruit flies (*Drosophila*).[11]

Many management decisions are characterized by a combination of uncertainty and intractability. For instance, the intractability of scheduling problems, as described previously, can go with the sudden occurrence of unexpected events whose consequences could not be foreseen—such as transportation routes being closed down owing to construction, earthquakes, or wars. Here, an adaptive toolbox of flexible heuristics is likely to be superior to long-term planning and utility maximization.

Nevertheless, if you read a book on management or take a course in a business school, you likely will not learn about the distinction between small and large worlds, or the fact that these need different tools to make good decisions. Similarly, most economic theories reduce large worlds to small worlds so utility maximization can be applied. That can be an interesting theoretical exercise, but it is of little help for practicing managers who have to make decisions under uncertainty. The blind spot for uncertainty can be traced to the classic writings in decision theory by Duncan Luce and Howard Raiffa, who distinguished risk and ambiguity (and unfortunately called the latter uncertainty).[12] That terminology eliminates all large worlds from the domain of decision science. In Tversky and Kahneman's influential paper "Judgment under Uncertainty: Heuristics and Biases," the term *uncertainty* is used even beyond ambiguity to encompass situations of risk.[13] Similarly, popular books in behavioral economics, such as *Freakonomics* and *Thinking: Fast and Slow*, talk about uncertainty and yet deal with either risk or ambiguity. This conceptual confusion leads to the mistaken impression that utility maximization applies to all problems.

What Does VUCA Really Mean?

In chapter 1, we used the popular term *VUCA*, a term that needs a precise definition. The concepts of volatility, uncertainty, complexity, and ambiguity are lumped together in this acronym, although they are not the same. They have fundamentally different meanings in small and large worlds. Assuming a small world of risk, finance theory defines *volatility* as the standard deviation of a variable over time, such as the performance of a stock. By that definition, volatility assumes stability over time. For instance, Markowitz's portfolio method assumes a situation of risk in which volatility can be precisely estimated from the past. In situations of uncertainty, however, volatility means more than random fluctuation; it refers to unforeseeable change and disruption. Here, fine-tuning the weights of assets based on past data can no longer produce an optimal portfolio in the future. Simple heuristics such as $1/N$, which assigns equal weights, can do better. In fact, in the real (large) world of financial uncertainty, $1/N$ has been shown to outperform Markowitz's Nobel Prize–winning method and to perform on par with the most sophisticated modern versions of it.[14] Exchange-traded funds such as index funds based on the Dow Jones Industrial Average, which are closely related to $1/N$, also consistently perform above managed funds.[15] Under uncertainty, fine-tuning on the basis of the past can be futile.

Similarly, as mentioned in this discussion, the term *ambiguity* in decision theory alludes to a small world with unknown probabilities. Yet that is not its meaning in the expression *VUCA*, where it appears to mean basically the same as uncertainty, emphasizing lack of clarity and certainty. As in the case of volatility, ambiguity has fundamentally different meanings in a small world and a large world. That adds to the confusion in the literature. Complexity also has multiple meanings: In computational complexity theory, for instance, it refers to tractability, but it can also refer to the complexity of models. For instance, small-world models such as cumulative prospect theory are typically relatively complex, whereas large-world models such as $1/N$ are typically simple.

In this book, we define the term VUCA *as a large world. The* V *concerns unexpected change over time, the* U *and* A *refer to aspects of uncertainty, and the* C *refers to intractability.*

The general point is that probability theory and optimization models are excellent tools in situations of risk, whereas heuristics are appropriate tools under uncertainty. Nevertheless, this distinction is often neglected, and the term *optimization* is used even for large-world problems. There is frequent talk of managers approximating the "optimal" solutions or making "suboptimal" decisions. Such statements ignore the fact that in situations of uncertainty, there is by definition no such thing as optimization, and calling a decision suboptimal likewise makes little sense when no mortal being can know the optimal decision.

Recall from chapter 1 that although Simon was trained in expected utility theory, he quickly realized that it was of little use in the real world of managerial decision making and that successful managers thus do not even try to incorporate it. Similarly, Selten always distinguished between rationality in well-defined games and useful heuristics in the real, uncertain world. Optimization and heuristics are neither antitheses nor rivals. They are appropriate tools for different situations, for small and large worlds, respectively.

Uncertainty Enables Progress

Many people think of uncertainty as something negative, to be avoided. This uncertainty-averse group includes most economists, behavioral economists, and others who construct risk models but avoid dealing with uncertainty. Unsurprisingly, companies have risk management departments, but not uncertainty management departments. Yet uncertainty is real and needs to be confronted. In 2003, Robert Lucas, one of the most distinguished macroeconomists, claimed that macroeconomics had succeeded in preventing economic depression.[16] Five years later, small-world theorizing led the world blindly into the largest financial crisis since the Great Depression. The preoccupation with studying small worlds creates theories that provide illusory certainty and assume a stable world in which nothing new can ever happen, meaning that they are of little use to practitioners. Thinking in terms of expected utility maximization, small worlds, and equilibrium models not only overlooks the possibility of crisis but also has even more striking properties. In a small world, there would be:

- No innovation
- No profit
- No trade
- No need for qualities such as intelligence, expertise, or intuition

By theorizing in terms of small worlds, innovation is unimaginable: All possible actions, future states, and their consequences are fixed and known. Nothing can change in unexpected ways, and innovation becomes impossible when everything is known. Moreover, as Knight pointed out long ago, in a world of risk, there is no profit to be made; the same conclusion was reached more recently with the efficient market hypothesis.[17] When all players know the future and no profit can be made, there is no incentive to trade with others (the no-trade theorem). Beyond that, qualities quintessential to humans such as intelligence, expertise, intuition, emotions, and trust are then of little use. Even worse, the evolved psychology of our brain is mistakenly identified as a source of cognitive illusions and irrationality.

In fact, the human brain has evolved to deal with the real world, a world that is largely uncertain and intractable and in which heuristics, intuition, trust, and emotions are indispensable for survival. We should look at uncertainty as something positive: without it, life would be endlessly tedious and, apart from calculation, no form of intelligence would be needed. Fortunately, adaptive heuristics help us navigate a world of uncertainty.

Advantages of Heuristics

What is a disadvantage in a small world can be an advantage in a large world. The term *fast-and-frugal heuristics* signifies three advantages that heuristics have in large worlds: they enable fast decision making on the basis of little information; they can be accurate not despite but because of their speed and frugality; and they are transparent, which means that they can be easily taught and understood. Table 2.2 summarizes these advantages, as well as

Table 2.2
Differences between small and large worlds

Concept	Small Worlds	Large Worlds
Speed–accuracy trade-off	A speed–accuracy trade-off exists: making decisions faster leads to less accurate decisions.	A speed–accuracy trade-off does not generally exist: taking more time does not necessarily lead to better decisions; a *reverse speed–accuracy trade-off* may exist such that faster decisions are more accurate.
Effort–accuracy trade-off	An effort–accuracy trade-off exists: using more effort to get more information leads to more accurate decisions.	An effort–accuracy trade-off does not generally exist: using more information may not increase accuracy of prediction (compared with fitting); a *less-is-more effect* can exist such that simple heuristics make more accurate predictions and decisions.
Transparency–accuracy trade-off	A transparency–accuracy trade-off exists: using more complex and therefore less transparent models leads to more accurate decisions.	A transparency–accuracy trade-off does not generally exist: simple heuristics can be both more accurate and transparent.

Three widely assumed trade-offs exist in small worlds. In contrast, in the real, large world, all three trade-offs do not exist at a general level. Sometimes the trade-offs are reversed, such as when faster decisions are more accurate and when less information is more effective.

three widely assumed trade-offs of using heuristics (e.g., that they are faster but less accurate). These trade-offs do not apply, at a general level, under conditions of uncertainty, as we show in the subsequent discussion.

Smart Heuristics Are Fast and Accurate

Many people assume that making decisions quickly increases the likelihood of making errors. The reason given is the supposedly general *speed–accuracy trade-off*: the faster a decision is made, the less accurate it will be. Various dual-system theories assume this trade-off by opposing a system 1, which is fast, heuristic, intuitive, unconscious, and often wrong, with a system 2, which is slow, logical, deliberate, conscious, and always right.[18] Yet there is clear evidence that fast decisions can be more accurate than slow decisions, heuristics can be used consciously, and the alignment of the features of the two supposedly opposing systems makes little sense.[19] For instance, the $1/N$ heuristic is used deliberately in portfolio construction, is fast, and nevertheless can outperform the Nobel Prize–winning mean–variance portfolio. In general, every heuristic can be used consciously or unconsciously and can be more or less accurate than deliberate logical thinking.

In an uncertain world, decisions do not necessarily improve when more time is available. Experts in particular are good at making fast *and* accurate decisions. Consider an experiment in which expert handball players stand, in uniform, in front of a video screen showing a professional handball game.[20] At some point, the video image is frozen, and they are asked what the player holding the ball should do. That could be a loop, a pass to the left, a goal shot, or something else. Many of these expert players rely on the following fast-and-frugal heuristic:

Fluency heuristic: Choose the first option that comes to mind.

This heuristic conflicts with the speed–accuracy trade-off, according to which more time is always better. Would these experts indeed pick a better option if they had more time? In the experiment, the players first gave an immediate response when the video image was frozen and then were given the opportunity to study it more carefully for forty-five seconds and generate further options. After that, they were again asked what they now thought would be the best action. This judgment based on further deliberation was, on average, inferior to their first, intuitive judgment. How can that be? Figure 2.1 shows the explanation of this striking effect. The first option that came to experts' minds was on average the best one, the second the second-best, and so on. When a decision has to be made quickly, inferior options do not even come to awareness and therefore cannot be chosen. In contrast, having a lot of time increases the danger of an inferior option overriding the first option.

The faster-is-better principle holds more generally for experts. For instance, expert golfers made more accurate putts when they had only up to three seconds rather than unlimited time.[21] Experienced firefighters have learned to make quick decisions that are better than decisions made after

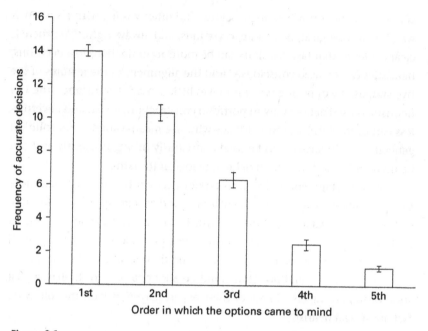

Figure 2.1

The first option that comes to mind is often the best one. Thus, relying on the fluency heuristic to choose the first option that comes to mind is not only fast but also accurate. The graph shows that the frequency of accurate decisions made by expert handball players decreases with the order in which the options came to mind. Note that relying on the heuristic works less well for novices. Error bars show standard errors. Based on Johnson and Raab (2003).

long deliberation.[22] Similarly, heuristics used by senior managers to decide which project to invest in have been found to be as accurate as their slower analytic methods.[23] And firms making faster strategic decisions often show both greater profits and more rapid growth.[24]

The fluency heuristic explains why faster decisions can be better. It also explains the situations in which this holds: a considerable amount of expertise is needed so the best option comes to mind first. The fluency heuristic leads experts toward good decisions, but not novices, who lack the necessary knowledge and experience. Expert athletes tend to trust their quick intuitions and appreciate their ability to make fast decisions. In business, by contrast, a culture of mistrust flourishes against fast and intuitive decisions that is first cultivated in business schools. The culture of many companies views slow decision making as a virtue. An executive can signal that they are a good decision maker by making decisions slowly. Thus, even when experienced executives have good intuitions, such as about which project to invest in, they often do not decide right away. They may instead engage in further deliberation, ask their staff to conduct lengthy analyses, or even hire expensive consultants to justify a decision that they had already made quickly and intuitively. This adds delays and costs; even worse, the process can result in replacing the first and better option with an inferior one. Negative error cultures, in which employees fear being punished for making errors and avoid taking risks for the company, exacerbate this tendency and lead to a slowing and even a complete avoidance of decision making (see chapter 11 for a more detailed discussion of this point). In contrast, the willingness to make fast decisions is more developed in family enterprises and fast-moving tech companies.

In sum, the speed–accuracy trade-off is not generally true in an uncertain world. In particular, experts relying on the fluency heuristic can make decisions that are both quick and accurate.

Smart Heuristics Are Frugal and Accurate

The second advantage of heuristics is their frugality. That is, they use little information, often as little as a single cue. A widespread, albeit incorrect account of why people use heuristics is the *effort–accuracy trade-off*: using heuristics reduces effort but decreases accuracy.[25] Such a trade-off is a general characteristic of situations of risk, but it does not apply to situations of uncertainty, where heuristics can save effort and simultaneously lead to more accurate decisions than would more effortful strategies. This striking benefit is called a *less-is-more effect*.

Consider companies seeking to predict which of their previous customers will continue making purchases. Experienced managers rely on a simple rule:

> *Hiatus heuristic: If a customer has not purchased within x months, the customer is classified as inactive, otherwise as active.*

In retail companies and airlines, the hiatus is often $x = 9$ months. Studies on twenty-four companies showed that future purchases are predicted more accurately by this heuristic than by machine-learning techniques (e.g., random forest) and complex marketing models that utilize additional predictor variables and computing power.[26] Managers use the hiatus heuristic not because they want to save effort at the cost of accuracy, but because the heuristic enables them to make more accurate decisions with less effort.

Why does the hiatus heuristic perform so well on the basis of a single cue? It is commonly assumed that more is better: the more data and computational power, the better the predictions. In situations of uncertainty, however, having more data is not always a good idea. Specifically, if one wants to predict the future but the future is not like the past, fine-tuning on the basis of the past leads to *overfitting*: that is, to projecting trends in past data onto the future, where they are no longer valid. Thus, when a company creates a complex model to predict future purchases by using massive amounts of customer data, it runs the risk of overfitting: The model succeeds at "explaining" past purchases but fails to predict future purchases.

For another example, consider predicting next week's flu-related doctor visits. For this task, engineers at Google developed a big data algorithm called Google Flu Trends (GFT). The idea was that if people experience flu symptoms, they are likely to conduct Google searches on flu-related terms; information regarding these searches should then help predict the spread of the flu much faster than any medical organization can. To develop the algorithm, the engineers analyzed about 50 million search terms, tested hundreds of millions of prediction models, and after selecting the best of these, made predictions of the proportion of flu-related doctor visits from 2007 to 2015. When the swine flu arrived out of season, beginning in March 2009 and peaking in October of the same year, GFT missed the outbreak. It consistently underestimated its spread, having learned from the years before that infection numbers were high in the winter and low in the summer (figure 2.2). In response, the algorithm was made more complex and the number of variables was jacked up from 45 to 160. This and further revisions did not improve the quality of predictions; in 2015, GFT was shut down.[27]

The flu happens in a dynamic, large world where viruses mutate and people enter search terms not only when they have symptoms but out of curiosity or for many other reasons. One way to avoid overfitting to the past is to use only the most recent data and ignore the rest. The *recency heuristic* relies on the most recent data point alone, in this case the rate of flu-related doctor visits in the last week.

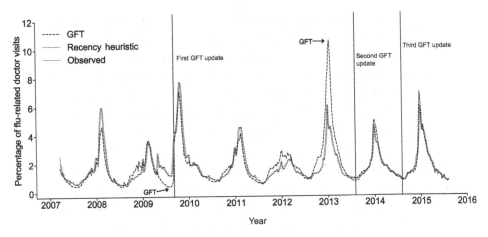

Figure 2.2
Predicting the weekly percentage of flu-related doctor visits using a single data point (the recency heuristic) reduces the prediction error by about half, compared with Google's big data algorithm Google Flu Trends (GFT). The mean absolute error for the recency heuristic is 0.20, and for GFT, 0.38. This holds for all updates of GFT and the entire time period from 2007 to 2015. For instance, when the swine flu broke out in the summer of 2009, GFT underestimated the percentage of flu-related doctor visits (dashed curve), whereas the recency heuristic (dotted curve) quickly adapted to the unexpected outbreak. The three vertical lines indicate the three GFT updates. The years signify the beginning of the year; that is, "2008" indicates January 1, 2008. Based on Katsikopoulos et al. (2022).

Recency heuristic: Predict that next week's rate of flu-related doctor visits will be the same as the most recent rate.

By relying solely on the most recent data point instead of big data, the heuristic can quickly adapt to out-of-season events due to mutation and is not sidetracked by irrelevant reasons for performing flu-related searches online. The recency heuristic predicted the flu consistently better for the eight years that GFT was tested, and it also outperformed all revisions of the big data algorithm.[28] Overall, it reduced the prediction error of GFT by about half (figure 2.2). In a volatile environment, one data point can lead to better predictions than big data.

The general lesson is this: to protect against overfitting the past, aim for simplicity. Simplicity means reducing the number of parameters of a model that need to be estimated from past data. The hiatus heuristic has only one free parameter, and the recency heuristic has no free parameter at all, making it robust in the sense that it cannot overfit. Less information is often more beneficial under conditions of uncertainty. Of course, this does not mean that ignoring all past information is best. Rather, it means that using only one or a few critical features, such as the hiatus, is an effective strategy. Under uncertainty, there is typically a ∩-shaped function between the number of features used and predictive accuracy.[29]

Smart Heuristics Are Transparent and Accurate

Transparency is a crucial feature of decision rules. A rule is transparent to a group of people if they can understand, memorize, teach, and execute it.[30] The hiatus heuristic, for instance, is transparent: a manager can easily understand, communicate, and apply it. In contrast, if a company acquires a complex machine learning technique such as random forest to predict future customer choice, the managers will not be able to understand how the predictions come about nor be able to explain its logic to others. The recency heuristic is likewise transparent, whereas the big data algorithm GFT is not.

There are two principal reasons for lack of transparency: complexity and secrecy. In the case of GFT, both apply: Google did not openly share sufficient details about GFT, such as the variables and algorithms used, perhaps because they wanted to keep the algorithms proprietary. Yet even if Google had openly shared this information, GFT would still be nontransparent to most people. The original algorithm was based on 45 search terms and was later increased to 160. Thus, revealing an algorithm does not guarantee transparency by itself.

Similarly, it has been widely assumed that transparent rules are always less accurate. In other words, to make the best decisions, one must rely on the most opaque rules. For instance, machine-learning researchers from the Defense Advanced Research Projects Agency have claimed that a general *transparency–accuracy* trade-off exists.[31] It has been illustrated by graphs such as figure 2.3.

As we show in this graph, this trade-off is not generally true. Although the GFT algorithm is less transparent than the recency heuristic, the latter is more accurate. Likewise, the hiatus heuristic, in spite of being more transparent, predicted future customer purchases more accurately than random forest, which constructs thousands of decision trees from previous customer

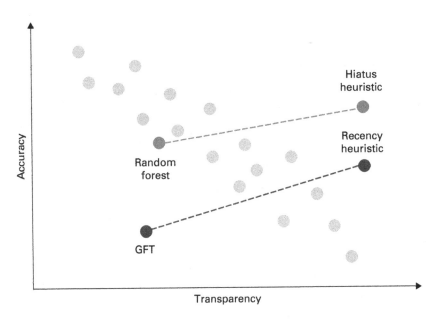

Figure 2.3
The transparency–accuracy trade-off does not generally hold. The light gray points illustrate the claim of a general transparency–accuracy trade-off in prediction: opaque algorithms predict best (top left), whereas transparent algorithms predict worse (bottom right). These points can be found in many sources but are rarely based on data. They suggest that transparency demands sacrificing accuracy. We have added counterexamples based on real data. The first pair shows that the transparent hiatus heuristic predicts customer purchases better than random forest, a complex and opaque machine-learning algorithm. The second pair illustrates that the recency heuristic predicts better than the nontransparent Google Flu Trends (GFT). Positions of the heuristics and the algorithms are relative and for illustration purposes only.

data and is one of the most powerful machine-learning techniques. These two examples, based on actual data, illustrate that there is no such thing as a general transparency–accuracy trade-off. Rather, we need to identify when more transparency is associated with higher accuracy and when it is not. That is the topic of the *ecological rationality* of heuristics, which we address in chapter 3.

The fact that there is no general transparency–accuracy trade-off is a positive result for *explainable artificial intelligence* (XAI), which has largely assumed this trade-off. For instance, most of the algorithms used to predict whether a customer will repurchase or someone accused of a crime will reoffend are

Table 2.3
Six common misconceptions about heuristics

Common Misconception	Clarification
Heuristics produce second-best results; optimization is always better.	In situations of uncertainty (e.g., business decisions) and intractability (e.g., chess), optimization is impossible. Here, heuristics are effective tools.
There are two systems of reasoning: system 1 is fast, heuristic, intuitive, unconscious, and often wrong; system 2 is slow, logical, deliberate, conscious, and correct.	This binary opposition is false. Heuristics can be used unconsciously or consciously and can lead to more successful decisions than can logical, deliberate thinking. And like heuristics, deliberate statistical thinking can be applied in wrong situations, as witnessed by the economic models that not only missed the financial crisis of 2008 but actually contributed to it.
Heuristics lead to biases; expected utility maximization does not.	This misconception follows from misconception 1. Because maximization is impossible in large worlds (uncertainty, intractability), the real issue is to know what heuristics to use in what situations. Using small-world tools such as maximization in large worlds can lead to illusions of certainty, as well as errors.
People rely or should rely on heuristics only in routine decisions of little importance.	Virtually all important problems involve uncertainty. Therefore, experts must rely on heuristics in situations with high stakes (e.g., investment, planning, and personnel decisions).
More data and computational power are always better.	This is true only in situations of risk. Good decisions under uncertainty require ignoring part of the available information to increase robustness and protect from overfitting.
Intuition should not be trusted; analysis is always better.	Without intuition, there would be little innovation or progress. The opposition between intuition and analysis is misguided; typically one needs both. Intuition is required to design a plan or to notice that something is going wrong, whereas analysis is required to evaluate the plan or locate the cause of a problem.

The misconceptions stem from the assumptions of small worlds, risk, and a "system 1" versus "system 2." Once large worlds and uncertainty are assumed and the value of different forms of thinking is appreciated, the misconceptions are resolved.

so complex that managers, defendants, and judges cannot understand how the predictions are made. To address this issue, XAI might try, for instance, to explain random forest in easy terms. However, that is difficult to do and risks distortion. Our approach offers a new solution: check first whether heuristics that are both transparent and accurate exist for the prediction task at hand before using difficult-to-explain, complex AI algorithms.

Common Misconceptions

In this chapter, we have introduced four main reasons for using heuristics: they are fast, frugal, accurate, and transparent. Heuristics make it possible to deal with large-world problems of uncertainty and intractability, where expected utility maximization and probability theory are unusable and even AI algorithms using big data struggle. The focus on small worlds and risk, rather than on large worlds and uncertainty, has generated a number of misconceptions about heuristics. They arise from the assumption of small worlds. Table 2.3 summarizes some of the most widespread ones.

From the next chapter on, we leave behind the unrealistic assumptions of small worlds and immerse ourselves more fully in the large world of uncertainty. Having examined the "why" of heuristics, we will learn about the "what" and "when": What are the different heuristics? And when or under what conditions do they work? The study of the adaptive toolbox answers the "what" question, and the study of ecological rationality the "when" question.

3 The Adaptive Toolbox

In a large world, there is no single best decision rule for all situations. Consider hiring, which involves much uncertainty about job candidates' future performance. Firms typically collect an array of information about the candidates, such as their education, personality, prior work experience, and social media activity, and factor all these elements into their hiring decision. Tesla founder and CEO Elon Musk, however, developed a very different approach. When Tesla was still a small company, Musk is reported to have used a heuristic that considers only a single cue.[1]

Musk's hiring rule: If a candidate has exceptional ability, make a job offer; otherwise, do not.

This rule is an instance of a *one-clever-cue heuristic*, a type of heuristic discussed in more detail in this chapter. Musk's rationale was that someone who has shown exceptional ability in the past is likely to show it again. Hiring can also be based on social heuristics, such as *word-of-mouth*. The Korean owner of a Chicago janitorial and cleaning company relied on his own employees to identify good candidates.[2]

Word-of-mouth: Ask existing employees for recommendations of suitable candidates.

The rationale of word-of-mouth is that employees tend to recommend people who they know are reliable because they feel responsible for the new hire, and their own reputation is at stake.

The set of rules, including heuristics, that an organization or a leader has at their disposal constitutes their *adaptive toolbox*. This toolbox metaphor is in direct contrast to theories that postulate only one general tool for all problems, such as expected utility maximization and Bayesian updating. As the saying goes, to a hammer, everything looks like a nail; to these all-purpose theories, everything looks like an optimization problem. The multiple tasks that exist in the real world, however, call for a diverse set of tools. The study of the adaptive toolbox thus addresses a *descriptive* question: What's in the toolbox?

To be a good decision maker, one must choose the right tool for the task at hand, similar to a builder who carries a toolbox and knows that a hammer goes with nails and a screwdriver with screws. This is the essence of *ecological rationality*: choose a tool that fits well with the demands of a task. For instance, Musk's hiring heuristic was an excellent choice when the company was small and in need of a vision for growth. At a later stage, with the company expanding, Tesla also needed employees who were good at routine work. Continuing to rely on exceptional ability alone would be counterproductive. Also, if fairness is a major concern in hiring, then the word-of-mouth heuristic used by the Korean owner may not be suitable, as the owner learned the hard way when he was sued for discrimination against non-Koreans. The study of ecological rationality thus addresses a *prescriptive* question: What heuristic should one use for a given task?

In general, the more experienced decision makers are, the more tools they have in their adaptive toolbox, and, importantly, the better their understanding of the tools' ecological rationality. Indeed, having a large repertoire of tools and being able to use them flexibly are hallmarks of intelligence. Let's now look at the major classes of heuristics in the adaptive toolbox.

Classes of Heuristics

Figure 3.1 lists five major classes of heuristics, along with specific examples in each class. These heuristics exploit core capacities of the human brain and recurrent features of physical and social environments. They have been studied extensively in the fast-and-frugal heuristics research program, and yet they are neither exhaustive nor in everyone's adaptive toolbox.

Recognition-Based Heuristics

Recognition is a core capacity of human memory and occurs with little conscious effort. Even the fact that one does not recognize an object can be informative. This is the rationale of the recognition heuristic.

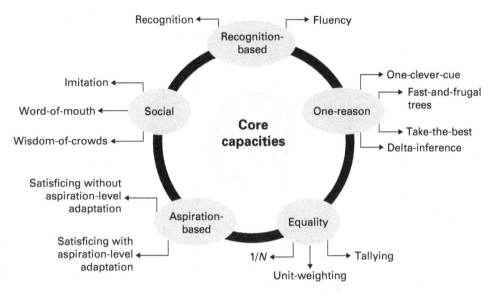

Figure 3.1
Major classes of heuristics in the adaptive toolbox and examples of each.

Recognition Heuristic

Consider a situation that involves choosing between two alternatives. It could be a company choosing one of two possible local banks in a foreign market or a consumer choosing between two brands of shoes.

Recognition heuristic: If one alternative is recognized and the other is not, then choose the recognized alternative.

The ingeniousness of this heuristic is that it exploits semi-ignorance, that is, the fact that one has heard of one alternative but not the other. The power of this rationale has been shown, for instance, in predicting winners of matches in the 2003 Wimbledon tennis tournament.[3] In the Gentlemen's Singles category, 128 players competed, resulting in 127 matches over seven rounds. To predict the winner of each match, one could use the official Association of Tennis Professionals (ATP) Champions Race Rankings or the ATP Entry Rankings, picking the player who is ranked higher (and who is usually also the higher seed in the tournament). The recognition heuristic, on the other hand, simply picks the player whose name is recognized. Tennis experts could not apply this heuristic because they recognized all the players; in contrast, amateur tennis players recognized only about

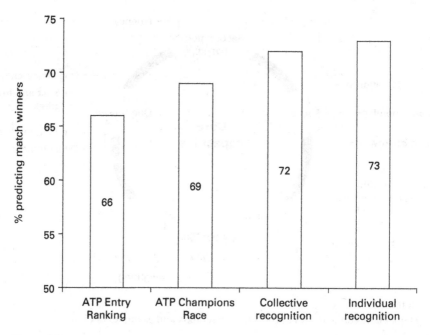

Figure 3.2
Amateur tennis players' recognition of the players predicted match winners in the
2003 Wimbledon tournament better than the official ATP rankings. ATP Entry Ranking
and ATP Champions Race are two different rankings of players; collective recognition
predicts that the player who ranks higher in name recognition among the amateurs
wins the match; individual recognition predicts winners according to the recognition
heuristic. Based on Serwe and Frings (2006).

half of the players and could apply the heuristic in about 40 percent of the
matches. In those matches, an amateur's recognition on average predicted
winners in 73 percent of the matches correctly, higher than the ATP rankings (figure 3.2). Because an individual cannot apply the recognition heuristic in all cases (i.e., when recognizing both or neither of the players), one
can alternatively rely on the recognition rates of players among the amateurs. This collective recognition led to a 72 percent predictive accuracy.

The recognition heuristic works here because recognition is highly correlated with the players' performance. Similarly, brand name recognition
is typically correlated with the quality of a product, and consumers rely
on brand names, preferring those they have heard of, in product selection.
When there are more than two products, brand name recognition is often

used to form a consideration set. Companies have tried to exploit consumers' reliance on recognition by investing in brand awareness rather than improving product quality. This tactic decreases the ecological rationality of using the heuristic for consumers because recognition then correlates more with the amount of advertising than with product quality.

Fluency Heuristic

The recognition heuristic relies on whether an alternative is recognized. The fluency heuristic, meanwhile, exploits the speed of recognition, choosing the alternative that is recognized faster. Therefore, it can be applied even if both alternatives are recognized. Fluency can also be used in situations where one has to generate options from memory, as in the case of handball players (see figure 2.1 in chapter 2). The fluency heuristic exploits the evolved ability of the human brain to detect subtle differences in recognition speed. Studies have reported that people can perceive the difference between recognition latencies exceeding 100 milliseconds.[4] As explained in chapter 2, years of experience make the fluency heuristic ecologically rational: the first option that comes to mind is often the best.

One-Reason Heuristics

Usually, there are multiple reasons for or against the available options. As a result, decision makers can be flustered by the sheer amount of information that they have to deal with. *One-reason heuristics* show that this does not have to be the case.

There are two types of one-reason heuristics. One type looks for a single clever reason and bases its decisions on it: the *one-clever-cue heuristics*. Musk's hiring heuristic is an example. The second may search for more reasons but also bases its decisions on only one reason. These are *sequential search heuristics*.

One-Clever-Cue Heuristics

A *clever cue* is one that is so powerful that considering other cues (or reasons) does not improve performance but rather slows decision making or even decreases performance. Consider the problem of how baseball outfielders catch a fly ball. One possible solution is that they calculate the trajectory of the ball and run to the point where it will hit the ground:

$$z(x) = x\left(\tan\alpha_0 + \frac{mg}{\beta v_0 \cos\alpha_0}\right) + \frac{m^2 g}{\beta^2}\ln\left(1 - \frac{\beta}{m}\frac{x}{v_0\cos\alpha_0}\right)$$

To calculate point $z(x) = 0$ where the ball hits the ground, the player would have to estimate both the initial angle α_0 of the ball's direction relative to the ground and the initial speed v_0 of the ball, know the ball's mass m and friction β, set the gravity acceleration g as 9.81 m/s^2, and be able to calculate tangent and cosine. Even then, the formula is overly simplified, in that it ignores wind and spin. Importantly, the true challenge is not computing the equation, but estimating its parameters, such as the initial angle and the initial speed.

Experienced players rely instead on simple heuristics. If the ball is high in the air, the gaze heuristic guides players to the ball.

> *Gaze heuristic: Fixate your eyes on the ball, run, and adjust your speed so that the angle of gaze remains constant.*

Figure 3.3 shows that by keeping the angle constant, the player arrives at the location where the ball lands. The angle of the gaze is a clever cue. Players who rely on it need not estimate the trajectory of the ball; in fact, they can safely ignore all the factors necessary for calculating the trajectory.

Figure 3.3
The gaze heuristic, a one-clever-cue heuristic, enables baseball players to catch a fly ball. To do so, the player adjusts the running speed so that the angle of the gaze remains constant. Various animals also use the heuristic to intercept prey and find mates. Source: Gigerenzer (2007).

The gaze heuristic was not invented by baseball outfielders. Bats, birds, fish, and other animals use it for hunting prey and finding mates.[5] The heuristic was also built into an extremely successful autonomous guided weapon: the AIM-9 Sidewinder short-range air-to-air missile.[6] The missile is an inexpensive but robust interception system whose "gaze" is directed at a point source of heat, which is the target. Although it was first employed in the 1950s, the AIM-9 Sidewinder is still in service in many countries, and new developments appear to be based on the same heuristic that maintains a constant angle of approach.

In the world of management, quite a few one-clever-cue heuristics can be found. Often, they are used to reject alternatives or to narrow down choices. Warren Buffett's famous rule, "Never invest in a business you cannot understand," specifies a single reason that is enough to exclude investing. Apple's strategic rule, "Only enter markets where we can be the best," is another case in point.

In their book *Simple Rules: How to Thrive in a Complex World*, the organizational scholars Donald Sull and Kathleen Eisenhardt described over 100 simple rules that people use in business strategy.[7] Many of the rules are of the one-clever-cue kind. For example, after the collapse of the Soviet Union, a Russian private equity firm used strategy rules when making its investment decisions, including "work only with executives who know criminals but aren't criminals themselves" and "invest in companies offering products a typical Russian family might purchase if they had an extra $100 to spend each month." The extent to which these rules are ecologically rational in other countries and times is open to investigation.

In cases where one clever cue is insufficient, a number of cues may be searched *sequentially*. Yet only one cue (reason) is used to make a decision. Fast-and-frugal trees, take-the-best, and delta-inference are examples of sequential search heuristics.

Fast-and-Frugal Trees

Emergency physicians must determine whether a patient needs immediate treatment or can be treated later; checkpoint soldiers must determine whether an approaching vehicle is friendly or carries a suicide bomb; and managers need to decide whether an employee should be promoted or not. Fast-and-frugal trees are tools for making such classification decisions. Unlike complex decision trees, a fast-and-frugal tree checks only a few cues or questions and tries to make a decision after each.

Fast-and-frugal tree: A simple decision tree with n cues and n + 1 exits.

It has three building blocks:

Search rule: Search through cues in a predetermined order.

Stopping rule: Stop the search if a cue leads to an exit.

Decision rule: Act according to what the exit specifies.

In an experiment, we asked managers to decide whether to keep or lay off a salesperson on the basis of their weekly sales performance.[8] The mean, trend, and variation of the performance were visible in a chart summarizing the sales record. A rule that many managers adopted is the fast-and-frugal tree shown in figure 3.4. First, consider whether the person's mean sales performance is above average. If yes, the person is not laid off and no other

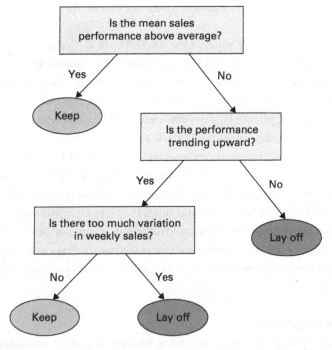

Figure 3.4
A fast-and-frugal tree used by managers to decide whether to keep or lay off a salesperson. If the mean sales performance is above average, the salesperson is kept. Otherwise, a second question about performance trend is asked, which may or may not lead to a decision. A third question about performance variation is asked in cases where the first two questions do not lead to a decision.

questions are asked. If the performance is below average, then the next question asked is whether the performance trend is upward. If not, the person is laid off; otherwise, a final question about the variation of sales is asked and a decision is then made. Unlike in a full decision tree, the order of cues is important in fast-and-frugal trees. The first cue can immediately lead to a decision, and the other cues cannot overturn that decision. For instance, a person with an above-average performance is kept even if the trend is downward and the weekly sales fluctuate widely.

Take-the-Best and Delta-Inference

Fast-and-frugal trees are heuristics for deciding on a single target (e.g., whether to fire an employee), whereas take-the-best and delta-inference are heuristics for choosing between two alternatives. Their logic and building blocks are otherwise similar to those of fast-and-frugal trees. The difference between the two is that take-the-best typically processes binary cues (e.g., whether a job candidate has a college degree), whereas delta-inference can process all types of cues, continuous, categorical, and binary (e.g., the candidates' IQ scores and their education levels). The *delta* in *delta-inference* refers to a threshold value above which the alternatives are judged to differ enough on a cue; this is when the search stops and a decision is made.

Consider the National Football League (NFL), the league for professional American football. In the US, it is the most popular sports league in terms of revenue generated, and NFL games are watched by millions every week throughout its playing season. The journalist Gregg Easterbrook used to write a football column called "Tuesday Morning Quarterback" for ESPN. In 2007, two readers wrote to him independently proposing a simple prediction model: the team with the better record wins; if the records are equal, then the home team wins.[9] In essence, this model is an example of delta-inference, in which the first cue is the teams' win–loss record and the second is home team or away (see figure 3.5). The delta in the "win–loss record" cue is set at 0 (i.e., any difference will lead to a prediction), and the home-team cue is binary.

This simple heuristic beat all but one of the dozens of experts whose records were tracked by Easterbrook in the 2007–2008 season. It achieved almost the same feat in the 2008–2009 season, beating all but two experts.[10] At times, Easterbrook questioned the picks made by the heuristic and replaced them with his own favorites. By doing so, the predictive accuracy was *lowered!* Using this heuristic, one does not need to have any insider information, spend time

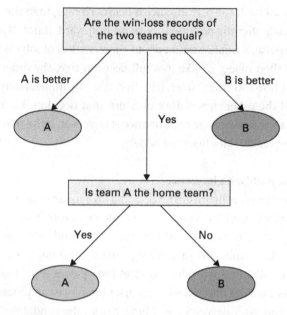

Figure 3.5
The delta-inference heuristic applied to predict winners of NFL games. Between two teams, the team with the better record is predicted to win the upcoming matchup. If the two teams' records are equal, then the home team is predicted to win.

reading reports and conducting sophisticated analyses, know the histories of the competing teams, or even understand the rules of American football. All the information needed can be found easily on any website publishing NFL game information.

One may attempt to improve the accuracy of delta-inference by trying to find the "optimal" deltas (i.e., deltas best fitted to past data). In a study of thirty-nine real-world problems—such as predicting which of two high schools would have a higher dropout rate and which diamond of a pair would sell for a higher price—we found that simply setting the deltas at 0 was just as accurate as using optimally fitted thresholds. The heuristic is also as good as complex models, such as Bayesian linear regression.[11]

Equality Heuristics

One-reason heuristics work well when there is a powerful cue. However, in situations where the cues are similarly informative, *equality heuristics* are the better choice. They integrate cues in a simple manner, such as by summing the reasons pro and con. This sets equality heuristics apart from optimization models that estimate weights for different reasons and take interdependencies and interactions among cues into consideration.

Tallying

Tallying is based on humans' core ability to count and compare numbers. It is a tool used to make classification decisions, and it works the same as fast-and-frugal trees but is based on the opposite logic. Instead of ordering cues and searching them sequentially, tallying treats all cues equally. Consider a task with n binary cues, where a *positive* cue value indicates category X and k ($1 < k \leq n$) is a classification threshold.

> *Tallying: Set a number* k. *If a target has* k *positive cue values or more, classify it as in category X; otherwise, do not.*

In essence, tallying embodies democratic voting among cues. It is simple and transparent and can lead to highly accurate classifications. For instance, the avalanche researchers Ian McCammon and Pascal Hägeli designed a tallying rule called the "obvious clues method" to evaluate avalanche risk: The situation is classified as dangerous if more than three of seven clues are present.[12] These clues, such as whether there has been an avalanche in the past forty-eight hours and whether there is liquid water on the snow surface as a result of recent sudden warming, were derived from years of observations and are indicative of avalanche risk. When tested against eight more complex methods, the obvious clues method achieved the highest prevention rate (i.e., accidents that would have been prevented). Allan Lichtman's *Keys to the White House* model, which predicts which candidate will win the popular vote in the US presidential election, is another example.[13] Since its first prediction in 1984, this tallying model has correctly picked all the winners with the exception of 2016, where it predicted that Donald Trump would win the popular vote (Trump won the presidency but not the popular vote).

Unit-Weighting

Organizations often use multiple linear regression to predict the values of a continuous variable, such as sales of a product. These models estimate the weights of cues to reflect their relative contributions. *Unit-weighting*, in contrast, weights all cues equally to reduce estimation error. At first glance, unit-weighting seems to be a good example of the effort–accuracy trade-off: by dispensing with the effort of estimating cue weights one ends up with lower judgment accuracy. However, a landmark study by the psychologists Robyn Dawes and Bernard Corrigan showed that this is not the case. In three of four tasks they examined, including predictions of college students' grade point averages, graduate students' academic success, and patients' psychiatric diagnoses, unit-weighting was more accurate than multiple linear regression. In light of this finding, Dawes and Corrigan proclaimed that to make good judgments, "the whole trick is to decide which variables to look at and then to know how to add."[14] Knowing the exact cue weights is of little value.

In assessing the personalities and attitudes of their employees, organizations often survey their potential or current employees using multiple-item scales that are weighted equally to form a composite score. Does it mean that answers to each item truly matter equally to the assessed underlying construct? Probably not. But there are two main reasons why unit-weighting is a good rule. First, the exact weighting scheme has little impact on the rankings of people being assessed. Second, the more items that are used, the larger the number of weights and correlations between items that need to be estimated, and the higher the estimation error. To avoid overfitting, it is thus reasonable to just weight the items equally.

1/N

Now consider a different type of problem: how to allocate limited resources to N alternatives, such as a limited budget to different divisions of a company or a limited amount of savings to different investment products. Once again, there are two visions for how to solve this problem. One is to get as much data as possible from the past, use the data to estimate weights for each alternative, and allocate resources according to the weights: that is, allocate more resources to those alternatives with greater weights. The other vision is for situations of uncertainty where the future is unlikely to be like the past. Here, one needs to simplify to avoid estimation error: that is, overfitting on the past. The $1/N$ heuristic allocates equal amounts to all alternatives and uses the principle of diversification in the same spirit as tallying and unit-weighting. As mentioned in chapter 1, Harry Markowitz's mean–variance model reflects the first vision, whereas the $1/N$ heuristic that he used for his own investments is in the spirit of the latter. Here, $1/N$ has been shown to perform on par with or better than the mean–variance and other highly sophisticated investment models, with considerably less time and effort.[15]

Beyond monetary allocations, the $1/N$ rule is also considered a fair way for parents or supervisors to distribute attention among children or employees. Interestingly, parents' use of the fair $1/N$ rule may wind up causing the *middle-born-child effect*: Growing up, middle-born children (e.g., the second child in a three-child family) receive fewer resources from their parents.[16] Assuming that parents allocate resources equally among their children at any given time, the firstborn will receive all the resources before others are born, and the child born last will get all the resources after the older ones become more independent or leave home. The middle ones never have such opportunities and must share the resources all the time. Therefore, cumulatively, they receive fewer resources, despite their parents' intent to be fair. This counterintuitive result exemplifies that the outcome of a heuristic depends on its environment (here, the number of siblings): if there are two children, the parents' goal of fairness is achieved, but not otherwise.

Aspiration-Based Heuristics

Kurt Lewin is often credited as the founder of social psychology. Among the countless discoveries that he made, one is the concept of aspiration, a goal that people are motivated to achieve. The concept was later borrowed by Herbert Simon and became the key ingredient of his well-known satisficing heuristic.

Satisficing

The heuristics introduced so far help one choose between two or a few alternatives. The *satisficing heuristic* can handle a large number of options, even in situations where one does not know how many alternatives exist. In its basic form, when evaluating options on just one attribute, such as price or expected profit, it has three steps:

Step 1: Set an aspiration level α and examine the options one by one.

Step 2: Choose the first option that satisfies α.

Step 3: If no option has satisfied α after time β, then change α by an amount γ and continue until a satisfying option is found.

If only the first two steps are used, the procedure is called *satisficing without aspiration-level adaptation*; if all three steps are used, it is *satisficing with*

aspiration-level adaptation. In business, satisficing is used for pricing commod-
ities. An analysis of over 600 German used car dealers showed that 97 percent
of them relied on satisficing, with or without aspiration-level adaptation. The
most frequent strategy was to begin with an average price, lower the price
after about four weeks, and repeat the procedure until the car was sold.[17]

The basic form of satisficing can be easily generalized to more than one
attribute by setting an aspiration level for each attribute. Suppose that a
venture capital firm wants to invest in a start-up in an emerging field and is
concerned with three attributes: the excellence of the company's five-year
vision, the proportion of engineers among all employees, and the charisma
of its founders. Using the satisficing heuristic, the firm sets an aspiration
level on each attribute, starts the search, and settles on the first start-up that
meets all the aspirations.

There may be better alternatives out there. But two factors besides uncertainty make satisficing a good rule: the cost of making a search and market competition. When search is a necessary part of the decision-making
process, it generally imposes a cost, as most people who have bought a house
can testify. The satisficing heuristic effectively sets a stopping rule on the
search and prevents the search cost from getting out of control. Moreover,
good things are desired by many; to get them usually involves competition.
If one keeps searching without making a decision, good opportunities will
likely be gone, picked up by others. Monetary investments, house buying,
and mate choices are all like that. Therefore, it is imperative for one to
know what one wants and act fast when a good alternative is obtainable.

The so-called secretary problem bears a resemblance to the two-step satisficing heuristic, but it assumes a small world where the number of options
n is known (and n is not very large, to avoid endless search). In this problem, a company aims to find the best secretary by interviewing candidates
one by one and deciding whether to give a candidate an offer immediately
after the interview. Once a candidate has been rejected, the company cannot
recall that candidate at a later point. When the total number of candidates is
known, the solution that maximizes the probability of getting the best secretary is to interview the first 37 percent of the candidates without making an
offer and then keep interviewing until a candidate with a higher quality is
found. However, if the number of candidates is unknown and the goal is to
select an excellent instead of the best secretary (e.g., top 10 percent), then a
simpler solution works better. It is called "Try a dozen," in which 37 percent

is replaced by a fixed number, 12. This heuristic has a higher chance of finding a suitable secretary with a substantially reduced search time.[18] Interestingly, after his first wife died, the astronomer Johannes Keppler is recorded to have considered eleven women as possible replacements before making his final choice. He had a productive second marriage with seven children, during which he wrote four more major works.

Social Heuristics

All the heuristics introduced so far in this chapter can be used to solve social and nonsocial problems. For instance, satisficing can be used to choose a house to buy, but also to choose a partner to marry. Yet there is another class of heuristics that are genuinely social, in that they rely on social information only. We introduce here three kinds of social heuristics: *imitation*, *word-of mouth*, and *wisdom-of-crowds*.

Imitation

Imitation is an enabler of human culture. No other species copies the behavior of others so generally and precisely as humans. From a very young age, children can already imitate the actions of others and understand their intentions, emulate the behaviors of adults and peers as a means to learn and affiliate with in-group members, and conform to majority behavior and social rules. Chimpanzees also imitate, but only occasionally and much less skillfully.[19] Learning by imitation not only helps children survive in an unfamiliar, uncertain, and possibly dangerous world but also instills stability in human groups and facilitates the passing of knowledge and social norms over generations.

Companies also frequently engage in the same kind of social learning by imitating other companies' successful products and technologies. For example, Amazon released Echo, an Internet-based home assistant device, in 2015. It was a huge success despite concerns over privacy. One year later, Google released a remarkably similar product called Google Home, and Apple did the same with its own Home Pod in 2017. Imitation provides a quick and relatively safe way for companies to enter markets. Instead of investing millions or billions in a cutting-edge technology whose commercial success is uncertain, companies can simply copy and improve on a market-tested idea, reducing the probability of failure. That said, blatant imitation without at least some attempt at differentiation (e.g., a lower price or new features) can hurt a copycat by putting it in a disadvantaged "late-mover" position as well as the entire market by preventing the invention of better technologies and products.

Word-of-Mouth

With the *word-of-mouth heuristic*, one's decisions are based on others' recommendations, as stated at the beginning of this chapter. Companies use it to find good hires and reliable business partners, job seekers to narrow down potential employers, and consumers to decide where to dine and what to buy. To be successful, word-of-mouth requires a relation of trust and long-term dependence between the person who asks and the one who recommends. It stops working when that trust is abused, especially when recommenders have other goals in mind than providing the most truthful information or suitable alternatives.

Wisdom-of-Crowds

In a short paper published in the journal *Nature*, Sir Francis Galton reported the first documented case of the wisdom-of-crowds.[20] Around 800 people bet on the weight of a dressed ox at a country fair in Plymouth, England. Galton gathered all the betting tickets and found that the mean of the estimates was only one pound off the actual weight.

> *Wisdom-of-crowds: Estimate a quantity by averaging the independent judgments of many people.*

The foundation for the wisdom-of-crowds is the law of large numbers in statistics: the larger a sample, the closer the mean of the sample to the true value. A key condition for the mean judgment to be accurate is that individual estimates are independent. If they are influenced by others—for example, a vocal leader—the estimates will not be independent and the mean can be biased, as in groupthink. In business, leaders too often utter their own opinions first, which affects what subordinates say (or even think) and makes wisdom-of-crowds no longer an ecologically rational heuristic. To avoid this pitfall, another heuristic can be useful: *first listen, then speak*. This heuristic is for the leaders, not the subordinates. It makes collecting the fruits of wisdom-of-crowds possible.

In the age of the Internet and social media, people increasingly rely on user ratings of restaurants, books, and many other products to make choices, hoping to harness the wisdom of crowds. If these ratings are made independently and without bias, then they will be good guidelines. Yet these conditions are not always in place. A 2021 report shows that among all the online reviews posted in 2020, 31 percent were estimated to be fake.[21] One source of the fake reviews is "bot farms," which manipulate rankings, stars, likes, and hearts for a fee.

Ecological Rationality of Heuristics

Could it be that Elon Musk makes better choices on the basis of a single reason than using many reasons, or an entire assessment center? The study of the ecological rationality of one-reason heuristics gives an answer to this question—and it is yes. One can prove that there are conditions under which relying on one reason is as good as or better than considering more information. The dominant-cue condition (discussed next) is one. Yet the study of ecological rationality also prescribes when other classes of heuristics are expected to be successful. We have already mentioned some of these conditions. Here, we focus on two general results. The first shows that the distribution of cue weights provides a guideline for choosing heuristics from the adaptive toolbox, and the second explains why simple heuristics can predict better than complex models in situations of uncertainty.

Dominant and Equal Cues

Cues drive both the absolute and the relative performance of heuristics. Generally, one-reason heuristics are ecologically rational in conditions where a dominant cue exists, whereas equality heuristics are ecologically rational when cues are equally valid. To understand why, let us consider the situation in which n binary cues are available to make a binary decision, such as hire or do not hire.

A linear model that weights and adds all cues has the form

$$y = w_1 x_1 + w_2 x_2 + \ldots + w_n x_n$$

where y is the criterion variable, x_i is the value of a cue i ($i = 1, \ldots, n$), and w_i is the decision weight of a cue that is ordered and reflects a cue's relative contribution after a cue or cues of higher ranks are considered. To simplify, all weights are positive. The model prescribes "hire" if y is positive; otherwise, "do not hire."

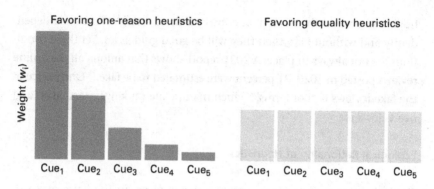

Figure 3.6

Distributions of cue weights (w_i) under which one-reason heuristics and equality heuristics are ecologically rational, respectively. Left: The dominant-cue condition that favors one-reason heuristics. Right: The equal-cue condition that favors equality heuristics. Based on Gigerenzer et al. (2022).

This linear model cannot make decisions more accurately than a one-clever-cue heuristic that bases its decisions solely on the most valid cue (i.e., Cue_1), if the sum of the weights of all other cues is smaller than the weight of Cue_1—thus, it is impossible for other cues to overturn the decisions made by Cue_1.[22] This is referred to as the *dominant-cue condition*, in which values of cue weights are such that

$$w_1 > \sum_{i=2}^{n} w_i$$

The left side of figure 3.6 shows an example of such a condition, in which the five cue weights are 1, 1/2, 1/4, 1/8, and 1/16. It is also an example of a stronger version of the dominant-cue condition where the weight of any cue is larger than the sum of the weights of subsequent cues. In this condition, it is guaranteed that one-reason sequential heuristics, such as take-the-best and fast-and-frugal trees, can never be outperformed by a linear model either.[23]

When all the cue weights are equal, as illustrated on the right side of figure 3.6, it is clear that one-reason heuristics cannot work better than equality heuristics such as tallying. In this equal-cue condition, no cue is better than any other; therefore, to make a good decision, one needs to take all cues into account. This is also the condition where no linear models that weight cues differentially can outperform tallying.

When cues are highly correlated, the dominant-cue condition is more likely to hold, as information added by other cues besides the most valid one is limited. In the aforementioned study that examined delta-inference in thirty-nine real-world problems, the top three cues in each problem tended to be highly correlated, and the dominant-cue condition held for most cases. That is the main reason why delta-inference with delta at 0, which decides almost exclusively on the basis of the most valid cue, did as well as linear regression across all problems. On the other side, when cues are independent, the equal-cue condition is more likely to hold. Although it is rare for cue weights to be precisely equal, equality heuristics can still be ecologically rational when cue weights do not differ much, or are difficult to estimate, because of the instability and uncertainty of the environment, the insufficiency of data, or both.

The Bias–Variance Dilemma

Take a look at figure 3.7. Two players threw darts at the board. Which player did better? Most would say player A. Yet this player has a clear bias: the darts are all to the lower-right side of the bull's-eye. Player B has no bias, as the average position of the darts is in the bull's-eye; however, the darts are all over the place and far from the target. This analogy helps explain why and when heuristics predict better than more complex models.

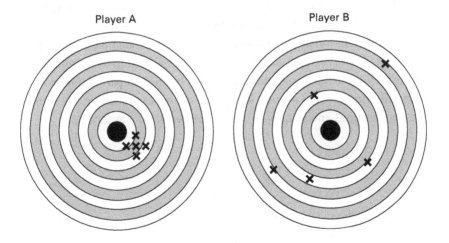

Figure 3.7
A dartboard illustration of the bias–variance dilemma. Player A's darts show a clear bias but only small variance, as the darts are all to the lower-right of the bull's-eye but close to each other. Player B's darts show no bias but considerable variance, as the average position of the darts is in the bull's-eye, but each dart is quite far from the others. Based on Gigerenzer et al. (2022).

The prediction error of a model comprises three components:

Prediction error $= bias^2 + variance + \varepsilon$

where *bias* is the systematic difference between a model's mean predictions and a true value, *variance* reflects the sensitivity of a model to sampling error, and ε is the irreducible error caused by random noise.[24] In predicting the sales of a product, for example, a model makes a prediction x_1 based on one random sample of observations, x_2 on another sample, and x_s on sample s. The difference between the mean of these predictions $\bar{x}$ and the true sales values μ is bias, and the variability of these predictions around $\bar{x}$ is variance.

In a stable world and with an ample amount of data, one may find a model that has both a small bias and a small variance. In an uncertain world and with limited observations, however, a bias–variance dilemma is usually present: Models with fewer free parameters tend to have smaller variance but larger bias than models with more free parameters, analogous to the contrast between the two dart players. Heuristics such as $1/N$, one-clever-cue, and take-the-best have none, one, or only a few parameters to estimate. Therefore, benefiting from the smaller variances, they often have lower prediction errors than highly parameterized models, such as multiple regression and Bayesian models. The advantage is even stronger in conditions where heuristics have the same bias as complex models, such as the dominant-cue condition for the one-clever-cue heuristics.

Ecological rationality goes hand in hand with the adaptive toolbox: good decision makers need to have both a repertoire of decision tools and the ability to pick a tool that can handle a task well. In the next part of this book, we use this insight to take a closer look at the heuristics in the adaptive toolbox of organizations and leaders, as well as their ecological rationality.

Part II

4 Hiring and Firing

When Coca-Cola hired Douglas Ivester to replace Roberto Goizueta as chairman and CEO, it seemed a safe bet: Ivester had been groomed by Goizueta as his successor and had excelled in his role as chief financial officer and number 2 at the company.[1] Soon into Ivester's tenure, however, it became clear that the company had made a big mistake: Ivester was a disaster as CEO. Although he was great with numbers and loved to do things in an orderly, "rational" manner, he lacked leadership, political, and people skills. Ivester antagonized powerful players within Coca-Cola, as well as key bottlers, a main constituency in the company's ecosystem. He handled crises poorly, responding slowly and lacking emotional intelligence. By the end of Goizueta's sixteen-year tenure as CEO, Coca-Cola's market value was three times higher than when he began, whereas it hardly moved during Ivester's stint, and earnings and return on shareholders' equity declined. Less than three years after he took the helm, Ivester was forced out by the company's major shareholders. He walked away with a golden parachute of at least $30 million.[2]

Ivester's case is by no means an exception. Both external hires and internal promotions often fail to live up to expectations, with some estimates suggesting that 50 percent of hires do not work out. Although these estimates are not exact, they demonstrate one thing: hiring errors are common, and as the case of Ivester illustrates, they can be costly. The costs of poor hiring decisions include

- Decreased productivity
- Lower team morale
- Termination costs such as golden parachutes and legal fees from unlawful termination
- Replacement costs

On the flip side of the costs of bad hiring are the benefits of hiring the right people. Employees—human capital—are the most fundamental resource of any organization. Employees make strategic decisions (chapter 5), drive innovation (chapter 6), negotiate deals and resolve conflicts (chapter 7), work together in teams to accomplish organizational tasks (chapter 8), and provide leadership (chapter 9). Being able to recruit outstanding employees is key for any successful organization.

Moreover, the implications of hiring decisions often extend beyond the organization and the hired individuals to society at large. For example, social injustice occurs when individuals from privileged backgrounds can access careers in prestigious companies and high-paying jobs thanks to their personal connections, but job seekers from less privileged backgrounds cannot. The same holds when starting salaries for women are lower than for men or when minorities are disadvantaged despite comparable qualifications.

Making the right hiring decisions is thus of utmost importance. Organizations apply various approaches to identify star performers and avoid bad hires. One such approach is the one-clever-cue hiring heuristic of Tesla's Elon Musk that we briefly encountered in chapter 3.

Elon Musk's Hiring Heuristic

Organizations often employ a "more-is-more" strategy for hiring decisions: they request a great deal of information about job applicants, including their education level, grades, reference letters, personality, intelligence test scores, and work experience.[3] That is not how Tesla CEO Elon Musk is reported to have approached hiring at a time when Tesla was still small and he was personally involved in hiring decisions.[4] When considering a job candidate, he used a one-clever-cue heuristic based on "evidence of exceptional ability": only if he found convincing evidence of exceptional ability did he hire the applicant. To judge whether an applicant had exceptional ability, he asked the following question: "Tell me about some of the most difficult problems you worked on and how you solved them." The rationale is that it takes exceptional ability to solve very difficult problems; moreover, it is difficult to fake a convincing answer to this question if probed for details, as Musk did when evaluating a candidate's response.

The approach to hiring taken by Musk is fast and frugal. It is fast because it dispenses with lengthy questionnaire batteries, time-consuming rounds

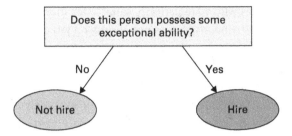

Figure 4.1
Elon Musk's one-clever-cue hiring heuristic, depicted as a simple tree with a single level and two branches. If a candidate provides evidence for exceptional ability, make a job offer; otherwise, do not.

of interviews, and expensive assessment centers. It is frugal because it relies instead on a single cue. The simplicity can be gleaned from figure 4.1, which represents the process as a simple tree. The tree has only one level with two branches, corresponding to the single cue that Musk used: exceptional ability.

When Does Musk's One-Clever-Cue Heuristic Work?

Musk's heuristic may be fast and frugal, but is it also effective? Or, more precisely, what are the ecological conditions under which it is expected to work well? As we have seen in chapter 3, one-clever-cue heuristics work well under the dominant-cue condition. A key enabler of this condition is redundancy among different cues. *Redundancy* here means that exceptional ability is likely correlated with other cues that predict future job performance, such as persistence, hard work, performance on intelligence tests, work samples, and experience.[5] Such redundancy implies that exceptional ability alone can be a powerful predictor because it captures to a considerable extent the information contained in the other cues. This suggests that making job offers based on evidence of exceptional ability may indeed be not only fast and frugal, but also effective.

The next question is whether any other factors influence the ecological rationality of this one-clever-cue heuristic. For one, the heuristic will work well only if exceptional ability (and its related abilities) is indeed required for the job. Exceptional ability was likely crucial when Musk started Tesla, in the face of humongous challenges. As Tesla grew, a broader variety of employees were needed, depending on the jobs in question. Continuing to rely on exceptional ability, however, would set a high bar and screen out a lot of "regular" candidates who were perfectly suitable for more routine jobs. Furthermore, because the heuristic does not consider a job applicant's interpersonal skills, it could result in hiring candidates who have exceptional abilities but are poor team players or even toxic employees.

Another important consideration is that for this heuristic to work, it needs to be possible to assess exceptional ability accurately. To the extent that a cue cannot be assessed accurately, the ability to predict future job performance is reduced. Musk addressed this issue by asking specific, probing questions. Doing so reduced the chance that applicants were faking their way through their answers. It also made it less likely that Musk's assessment would be influenced by demographic variables such as the applicant's gender, ethnicity, or age, thereby avoiding discrimination.

Jeff Bezos's Fast-and-Frugal Tree for Hiring

Like Musk, Jeff Bezos, the founder and CEO of Amazon, expected exceptional ability when still hiring applicants himself.[6] However, in addition, he looked for two more features: admiration and effectiveness. More specifically, Bezos first evaluated whether applicants had exceptional ability; if not, they were not hired. For an applicant who did, he considered a second question: Would he admire this person? If not, the applicant was not hired, as Bezos believed that he could learn from colleagues he admired. Third, he took into account whether the person would raise the average level of effectiveness of the group they would be joining, to ensure a steady increase in the level of performance in the company. Only if all three questions were answered affirmatively was the candidate hired.

Bezos's heuristic is more complex than Musk's. It can be captured as a fast-and-frugal tree with three levels, as displayed in figure 4.2. This diagram shows that Bezos's hiring strategy is noncompensatory. That means that only one cue is considered at a time, and lower-level cues in the tree cannot compensate for higher-level cues. For example, even if Bezos believed that a candidate would raise the average level of performance significantly, this could not compensate for a lack of exceptional ability or admiration, which would already have led to rejecting the candidate.

Figures 4.1 and 4.2 illustrate an important aspect of modeling heuristic decisions: a heuristic can function as a building block of another heuristic. In the next section, we show how to design fast-and-frugal trees so they

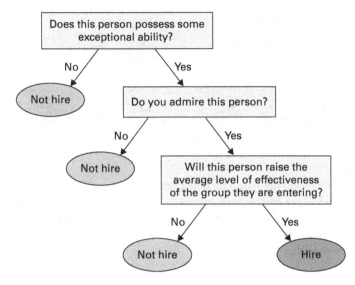

Figure 4.2
Jeff Bezos's hiring heuristic depicted as a fast-and-frugal tree with three levels. One cue is considered at a time, starting with exceptional ability. Positive values of lower-level cues cannot compensate for negative values of higher-level cues.

create a desired balance between rejecting qualified candidates and accepting unqualified candidates.

Flexible Fast-and-Frugal Trees

As we have seen in chapter 3, a fast-and-frugal tree, like many heuristics, has three rules: a search rule, a stopping rule, and a decision rule. In Bezos's example, the search rule is to search through the three cues sequentially, beginning with exceptional ability, then admiration, and finally effectiveness. The stopping and decision rules are to stop the search whenever a cue leads to a "not hire" decision and reject the candidate—unless all three cues are affirmative, in which case the candidate is offered the job.

Bezos's tree is only one of several possible fast-and-frugal trees using the three cues. In fact, given three cues, four different trees can be constructed that use the same cue order, and even more can be created if the cue order is allowed to vary, pointing to the flexible nature of this decision heuristic. These four fast-and-frugal trees are shown in figure 4.3.

The tree on the far left represents Bezos's hiring strategy. This tree is the most conservative, as it requires three affirmative cues for a "hire" decision

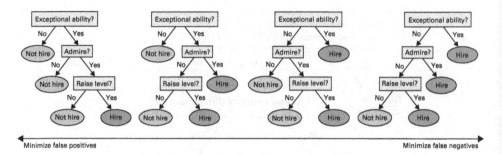

Figure 4.3
Four possible fast-and-frugal trees for hiring based on three cues. The leftmost tree represents Bezos's hiring strategy from figure 4.2. This tree minimizes false positives, that is, hiring an unsuitable candidate. In contrast, the rightmost tree minimizes false negatives, that is, not hiring a suitable candidate. The two in the middle balance the two potential errors. Based on Gigerenzer et al. (2022).

to be made. By setting the bar very high, this tree *reduces false positives*, that is, making offers to unsuitable candidates. At the same time, the tree *increases false negatives*, that is, rejecting candidates who would have been suitable for the job. The tree on the far right, by contrast, is the most liberal, as it makes a decision to hire if any of the three cues is affirmative. As a consequence, this tree reduces false negatives but increases false positives. The two trees in the middle of figure 4.3 balance the two errors of false positives and false negatives.

The concept of false positives and false negatives comes from signal detection theory and applies broadly to classification decisions, such as whether to hire someone or not or whether a person is sick or not after a positive test result. It illustrates an important insight: heuristics are not intrinsically good or bad; they need to fit the task environment, which includes the goals of decision makers. If an organization wants to minimize false positives, it should use a more conservative fast-and-frugal tree; if it wants to minimize false negatives, it should use a more liberal tree. Context also matters: in legal and cultural environments where it is difficult to fire employees, avoiding false positives is more critical, whereas in "hire and fire" cultures, avoiding false negatives is relatively more important.

Multiple-Hurdle Selection
Fast-and-frugal trees can also be used to design multiple-hurdle selection processes in which organizations screen out a proportion of applicants at each

step ("hurdle") and accept only those who pass all hurdles.[7] For example, in the first step, applicants may be screened through their submitted materials, in a second step through standardized tests, in a third step through assessment centers, and in a final step through an interview. This selection procedure can be modeled as a tree with the basic structure (but different cues) as in the leftmost tree in figure 4.3, which accepts only those who pass all the steps. The approach is, as we have seen, conservative: it reduces false positives but at the risk of screening out candidates who would have been suitable for the job.

Multiple-hurdle selection processes are noncompensatory, meaning that a candidate's strong performance on later hurdles cannot compensate for poor performance on earlier hurdles. Instead of assessing all these cues from all applicants, they save a considerable amount of time and other resources. Multiple-hurdle procedures are ecologically rational especially when there are a lot of applicants to choose from and when the early hurdles are inexpensive and easy to implement, thus enabling an organization to narrow down a large pool of applicants quickly and at low cost.

Choosing between Two Job Applicants with the Delta-Inference Heuristic

Musk's and Bezos's heuristics were intended for deciding about one applicant at a time. In other situations, organizations try to determine who the better of two applicants is. To do so, they can use the delta-inference heuristic. You might recall from chapter 3 that delta-inference makes decisions between two alternatives by searching through cues in order of their validity and stopping the search when the first cue discriminates between the two options. In a hiring context, for example, an organization might use the following three cues in sequence: job applicants' general mental ability, their conscientiousness, and ratings from structured interviews. These cues have been reported to be among the best predictors of future job performance across a broad range of jobs.[8] If the two applicants differ by at least a prespecified amount (i.e., a threshold of delta) on general mental ability, the applicant scoring higher is offered the job; if not (i.e., $\leq$ delta), conscientiousness is considered next, and so on.

Using real-world data of 236 applicants who were hired by an airline company, we examined how well managers can pick the better applicant of a pair by relying on delta-inference. Each applicant was measured on the three cues mentioned here; in addition, because all were hired, we knew their job performance three months later as rated by their supervisors.[9] There were 50,334 pairs of applicants who had different performance scores. From these pairs, we then drew small, moderate, and large random samples to simulate conditions of scarce, moderate, and ample opportunities for a manager to learn the parameters of the heuristic: the order of the cues and the delta of each cue. We compared the accuracy of delta-inference with that of logistic regression, a standard technique that always used all three cues to make a selection decision.

Figure 4.4 shows that if managers use delta-inference, they can select the better applicant more frequently than if they rely on logistic regression. This less-is-more effect held in all learning conditions, especially when learning opportunities were scarce. Moreover, using delta-inference allows quite frugal decision making, using on average fewer than half of the available cues. Besides illustrating the practical usefulness of delta-inference, these results demonstrate once again that the presumed "speed–accuracy," "effort–accuracy," and "transparency–accuracy" trade-offs do not generally hold in conditions of uncertainty (see chapter 2): relying on the delta-inference heuristic, managers can make decisions faster, more frugally, and in a more transparent way, while at the same time increasing accuracy.

As can be seen from figure 4.4, performance increases with learning opportunities. However, even with ample opportunities and using delta-inference, managers would be able to select the better candidate only 63 percent of the time. Predicting the future performance of job applicants is difficult, and despite relying on smart heuristics, errors remain common.[10]

Having established its performance, we next wanted to find out whether managers use delta-inference to make selection decisions, and whether they use the heuristic adaptively. We recruited human resource managers and business students with varying levels of experience in personnel decisions to participate in an experiment with two tasks: to hire a receptionist and to hire a data analyst. As shown in figure 4.5, both the less and the more experienced managers commonly decided with delta-inference, but the latter did so more frequently. As an indication of adaptiveness, the use of delta-inference increased when one cue was judged as far more important than the other cues, that is, when the dominant-cue condition introduced in chapter 3 held. This understanding of ecological rationality was particularly strong for more experienced managers. Thus, consistent with

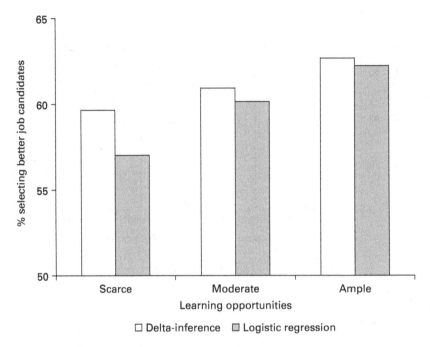

Figure 4.4
One-reason decision making (delta-inference) selected job candidates better than many-reason decision making (logistic regression). The advantage held regardless of whether learning opportunities were scarce, moderate, or ample (with random samples of size 30, 100, and 1,000, respectively). Based on Luan, Reb, and Gigerenzer (2019).

findings in other areas,[11] experienced managers were more likely to use a heuristic and to do so adaptively.

Social Heuristics for Hiring

Organizations often rely on the social heuristics of imitation and word-of-mouth to identify suitable job candidates. For example, when searching for executives, Fortune 500 firms tend to hire from companies that in the past had sent a large number of executives to other Fortune 500 firms.[12] The benefit of this imitate-the-majority heuristic is that it reduces the substantial uncertainty in top management hires and also speeds up the search.

In the word-of-mouth heuristic, organizations rely on their employees to recommend candidates for open positions. Evidence suggests that such

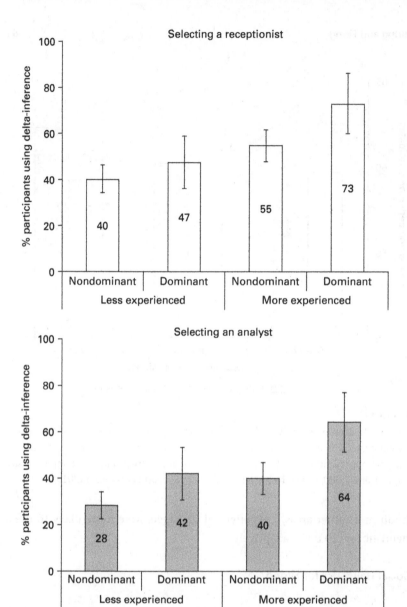

Figure 4.5

Experienced managers were more likely to rely on one-reason decision making (delta-inference) than less experienced managers. They were also more likely to use the heuristic when it was ecologically rational: that is, when the dominant-cue condition held. In the study, participants indicated which of two job candidates they would like to hire on the basis of three cues and did so for multiple comparisons. Participants who had made more than four personnel decisions (i.e., the median) in the past were categorized as "more experienced." Participants rated the importance of each of the three cues. In the "dominant" condition, the most important cue's rating was larger than the sum of the ratings of the other two cues. Error bars indicate standard errors. Based on Luan et al. (2019).

referral hiring works. For example, an analysis of twenty years of labor and social security records in the Munich, Germany, metropolitan area found that workers hired through referrals better fit the needs of the hiring companies and were less likely to leave.[13] Word-of-mouth can provide access to information that is otherwise difficult to get, thereby reducing information deficiencies in the labor market. In another study, employees referred higher-quality candidates when they received rewards for their referrals' performance, and high-ability employees recommended candidates of higher ability than did low-ability employees.[14] However, organizations also need to be wary of unintended negative side effects of using the word-of-mouth heuristic for hiring. Even though the US Court of Appeals for the Seventh Circuit ruled that word-of-mouth hiring was not discriminatory in the case of the Korean business owner mentioned in chapter 3, but rather the least expensive and most effective way of recruiting,[15] there is still the possibility that the practice can reduce employee diversity and risk discrimination. This illustrates how every heuristic—like any algorithm—has limitations and needs to be applied judiciously.

Transparent Decisions Reduce Discrimination

Concerns about discrimination are not limited to referral hiring but also apply to artificial intelligence (AI) algorithms used in hiring. As the Equal Employment Opportunity Commission (EEOC) chair Charlotte Burrows warned: "New technologies should not become new ways to discriminate."[16] To underscore this concern, on May 12, 2022, the US Justice Department and EEOC jointly released a warning that organizations' use of AI algorithms as part of hiring could lead to disability discrimination and violate the Americans with Disabilities Act.

The EEOC's technical assistance document issued on the same day offers some examples of how AI may discriminate against individuals with disabilities. For instance, organizations increasingly use AI chatbots to interact with applicants, and the algorithms underlying such chatbots might reject any applicant who reveals a significant employment history gap during this interaction. However, this gap may have been caused by a disability (e.g., by the need to undergo treatment), in which case the rejection constitutes a discriminatory decision. As the EEOC points out, such discrimination may well occur even when employers and software vendors argue that the decision-making algorithms they use are "bias-free" and have no adverse impact based on race, sex, national origin, color, or religion.[17]

To make matters worse, the AI algorithms that organizations increasingly use for hiring, promoting, firing, and other personnel decisions are often not transparent, making it difficult to determine in each case whether a decision was made fairly or whether discrimination occurred. This may be convenient for organizations as a way of trying to avoid responsibility. However, it does not do justice to applicants. Smart heuristics enable organizations to make not only accurate but also fair decisions. Their big advantage over more complex processes, including AI, is that their simplicity encourages transparency. And transparency encourages fairness because a decision process that is clearly unfair invites criticism and resistance. Thus, even though not all heuristics start off being fair, they make it easier to determine the source and size of the problem, enabling organizations to take countermeasures and make improvements over time.

As an example, a manager's simple rule to hire only men (or women) would be blatantly discriminatory for most jobs. However, one of the advantages of the adaptive toolbox is that it contains many heuristics, designed for specific purposes. If a company wanted to hire more minorities to create a more diverse workplace, it could consider using a variation of the $1/N$ heuristic.[18] In other words, it could set certain percentages or quotas of new hires to come from different groups and then try to identify and hire the best applicants within each group. Such quota systems are increasingly used in companies. Critics argue that under such systems, the most qualified applicants are not always hired. But this criticism misses an important point: the goal of the heuristic is not solely to hire the best person. Rather, the heuristic tries to balance two goals: hiring suitable people while also increasing diversity.

Are More Interviewers Always Better?

If more is better, as the traditional view posits, then you might think that having more interviewers would lead to better hiring decisions. When evaluating interviewees, managers often vote independently and apply the majority rule to decide who gets an offer. However, in such situations, if the interviewer with the best track record conducts the first interview, adding a second interviewer *never* increases accuracy.[19]

Consider a business that needs to identify the top ten candidates among a large pool of applicants. Assume that the best interviewer has a hit rate of

80 percent, meaning that they identify eight of the ten target candidates in the pool correctly but miss two (see figure 4.6, top). Adding a second interviewer with a hit rate of 60 percent (figure 4.6, middle) and applying the majority rule results in an expected collective hit rate of only 70 percent. This can be seen from figure 4.6 (bottom), where the votes of both interviewers are added. Four candidates get two votes, and they are among the targets. From the twelve candidates with one vote each, six are randomly selected, resulting in an expected number of three targets. Together, the expected performance of using two interviewers is $4 + 3 = 7$ correct identifications—one fewer than the best interviewer would have found if deciding alone.

One might need an additional six or more interviewers (with hit rates between 50 and 100 percent of the best interviewer's hit rate) to improve upon having the best interviewer alone. Again, less can be more. Under quite general conditions—independent votes and majority rule—adding a second interviewer to the best interviewer cannot improve decision making; it only makes it worse. The general lesson for a business is to invest in excellent interviewers, rather than trust in the collective decisions of a group of less-skilled persons. This lesson parallels situations where relying on one cue can be better than relying on many, as formalized by the dominant-cue condition and the bias–variance dilemma.

One reason that expert interviewers are better at identifying good hires is that they tend to use structured interviews. In structured interviews, applicants are probed on the same issues, just as Musk probed for exceptional ability in all candidates, greatly increasing the consistency of the cues assessed. These cues can then be processed using smart heuristics with clearly defined and consistently applied search, stopping, and decision rules. The validity of structured interviews is among the highest of all selection methods, much higher than that of unstructured interviews, which suffer from poor reliability.[20]

"Debiasing" Hiring Decisions?

Hiring managers' "stubborn reliance" on intuitive judgment has been blamed for poor hiring decisions.[21] According to this account, if practitioners used analysis instead of intuition, they could avoid hiring biases and make better decisions. In response to such concerns, some organizations have made an effort to "unbias" their hiring decisions and performance evaluations.

For example, Google developed Project Unbias to reduce unconscious decision biases.[22] This project includes a number of useful tools, such as

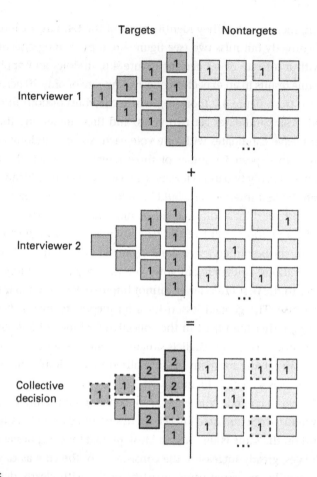

Figure 4.6
One interviewer can be better than two. Interviewer 1 has a hit rate of 80 percent—that is, identifies eight of the top ten candidates (targets) in a large pool of applicants correctly—and interviewer 2 has a hit rate of 60 percent. Even if interviewer 2 identifies the remaining two candidates that interviewer 1 missed, their collective decision (using majority rule) has an expected hit rate of only 70 percent, which is inferior to relying on interviewer 1 alone. Each box represents a candidate, and the top ten candidates are dark gray on the left side. The numbers represent the votes for each candidate. In the case of ties, candidates are selected randomly, as exemplified by the dotted boxes. Based on Fifić and Gigerenzer (2014).

checklists for interviews and performance appraisals. Checklists are help-ful because they direct attention to the most important cues. They can be effectively combined with sequential heuristics that use cues in order of their importance or validity, such as fast-and-frugal trees.

Unfortunately, such programs are still largely premised on the misguided idea that system 1 is to blame for biases in selection decisions and the prob-lem can be solved by getting decision makers to use more information and process it using system 2.[23] This ignores an important characteristic: hiring decisions are made under uncertainty. Under uncertainty, as we have seen before, heuristics help organizations with a very challenging task: to predict the future performance of job applicants. The problem with hiring decisions (and personnel decisions more generally) is not the use of heuristics, but the failure to systematically study the quality of interviewers and of cues and then use this information to design ecologically rational heuristics for hiring (and firing). Managers often rely on heuristics but without investigating and fully understanding which heuristics work under what conditions and why. As we have seen, complex quantitative models do not necessarily improve hiring decisions, as they are too fragile in this situation of uncertainty. Smart heuristics present an effective solution, combining simplicity with accuracy and transparency.

Once again, the problem is not simply in the minds of the people who make hiring decisions; it is also in the lack of systematic learning in organizations. Consider the following hiring paradox: organizations spend huge amounts of money and time on hiring but invest very little to find out whether their hiring processes are effective.[24] This is puzzling given the many hiring failures that occur. By systematically evaluating the effectiveness of their hiring processes, organizations could improve their adaptive toolbox of hiring heuristics, ultimately hiring more qualified candidates and rejecting more unqualified ones.

Smart Heuristics for Performance Management

Let us take a look at another type of personnel decision: performance man-agement, particularly promoting and firing employees.

Promoting and Firing Using Fast-and-Frugal Trees

Should an employee get a bonus? Be promoted? Or fired? Organizations make these decisions as part of what is called *performance management*. Do smart heuristics play a role in these decisions? To find out, we examined whether decisions to promote or fire employees are better modeled by weighting-and-adding, compensatory logistic regression, or by lexicographic, noncompensatory fast-and-frugal trees.[25] We gave decision makers performance profiles that varied on three performance-related cues for each employee: their average (mean) performance over the last half year, performance variation (i.e., random, unsystematic changes over time), and performance trend (i.e., systematic changes over time, such as upward or downward trends).

We found that fast-and-frugal trees were widely used, and even more so by experienced managers, of whom two-thirds relied on them. The majority of participants also adopted key features of fast-and-frugal trees adaptively in response to manipulation of the required distributions of positive (bonus) or negative (termination) decisions, consistent with the ecological rationality principle.

Stack Ranking

In our study, we required participants to lay off or give bonuses to a certain percentage of employees. Our approach was adapted from the so-called forced distribution or stack ranking performance management systems. Also known as *rank and yank*, this approach became (in)famous after General Electric (GE) CEO Jack Welch introduced the "20/70/10 split" rule: the top 20 percent, as ranked by their managers, were rewarded and the bottom 10 percent were fired.[26] The intention of this simple rule was to reward doers and remove underperformers, and the rule seems to have worked well for GE when Welch took over, at a time when the company had much deadwood.

However, the effectiveness of and reactions to such forced distribution ranking rules vary considerably.[27] And the criticism is predictable: how can such a simple rule do justice to the specific and unique situation of each employee? What this criticism misses is that no decision strategy under uncertainty is error-free; there are also disadvantages to implementing a more ambiguous performance management system that does not clearly specify who is rewarded and who is punished. As with all heuristics, stack ranking works well for only a specific purpose (trimming an organization) and situation (the existence of deadwood). Once its goal is achieved or the situation has changed, continuing to rank and yank forces managers to fire capable employees, making firms less functional. This may explain why the heuristic did not work well at Microsoft and may have even contributed to its decline in the 2000s.

Moving Forward

Although heuristics are widely used in making personnel decisions, few of them have been researched from the perspective of ecological rationality. Instead, the majority of studies have associated heuristics with biases and argued that decisions should be made via system 2 rather than system 1—despite the lack of evidence of such a duality.[28] We hope to change this narrative. To make good decisions under uncertainty, one needs to rely on both intuition and analysis. Smart heuristics provide a way to integrate these two. In so doing, heuristics can help make personnel decisions more transparent, more consistent, more fair, and more effective.

5   Strategy

Masayoshi Son was the richest person in Japan in 2021 and one of the most successful venture investors in the world. Born in Japan to poor Korean immigrant parents, Son founded the SoftBank Group, which has invested in business ventures such as Yahoo!, Alibaba, Uber, and ARM Holdings. The investments have been so successful that SoftBank was once the second-largest publicly traded company in Japan (after Toyota). To expand his investment portfolio and further increase global competitiveness, Son raised $100 billion to start the largest investment fund to date (the Vision Fund). Although some of his investments have failed rather spectacularly (e.g., WeWork), Son seems to have a knack for seeing things coming before others do, investing in companies that are more likely than not to become highly profitable. Where does his success come from?

Replicate a Successful Business Model in a New Market

SoftBank's annual report in 2000 gives a clue. In the report's section titled "Strategy," it says that SoftBank was pursuing a *time machine management strategy* that "fosters the global incubation of superior business models found through its venture capital operations in the United States."[1] An analysis of SoftBank's investment pattern shows that the strategy works like this: Son believes that technology-based business models such as e-commerce, social media, and ride-sharing services develop in different stages in different parts of the world. Typically, an impactful model gets its first success in the US, then spreads to other developed countries such as Japan and South Korea, and finally blossoms in countries such as China and India. To benefit from the developmental lags, SoftBank bases its selection of promising business models on its market analysis in the US and then incubates companies running on similar models in other countries. The process is like traveling in a time machine from the US to these countries to see the growth of a business once again.

To explain why SoftBank invested in Yahoo! Japan, for instance, Son said, "In the early stages of Internet in Japan, many said that Japanese and Americans are different. There are 10 reasons why Japanese Internet is not taking off. I said none of them are right; it's just a time lag. And, of course, Japanese Internet took off."[2] Yahoo! Japan followed the same model as Yahoo! by being a web portal that directs traffic and provides a multitude of services, such as news, auctions, and finance. It has been the most visited Japanese website for two decades, and in 2021, it was the second most popular search engine after Google in Japan. SoftBank's hugely successful investment in China's Alibaba, modeled after Amazon, is another example. Core to the time machine management strategy is imitation.

Time machine heuristic: Imitate a business model that has been successful in the United States and replicate it in other markets.

The heuristic has worked well not only for SoftBank but for many other information technology (IT) companies. One of the most infamous cases is Rocket Internet, founded by three German brothers, Marc, Alexander, and Oliver Samwer. While interning in Silicon Valley in 1998, the brothers witnessed the growing popularity of eBay and decided to set up a similar online auction site in Germany. Their website, Alando, is a direct copy of eBay, down to the color scheme of its logo. Alando was an instant success, getting more than three million page views in the first month it went live. The brothers soon sold the website to eBay for $43 million. After this initial taste of success, they continued to clone other Internet companies successful in the US, such as YouTube and Facebook, set them up in markets the US companies had not entered, and sell the clones back to these companies for huge profits.

Although the Samwer brothers' business strategy was indisputably successful, it also got plenty of criticism from the innovation-driven entrepreneurship community. In an interview with *Wired* magazine, the brothers insisted that their version of the time machine heuristic might look simple, but for it to work, careful execution was needed.[3] To develop the imitated business model in a new market, one must act fast before others do, invest heavily at the beginning to make the business operational, and adapt the model to the intricacies of the local culture and regulations. Indeed, these are the key conditions for the time machine heuristic to be ecologically rational. A late entrance without the backing of sufficient capital and a good understanding of the new market would doom a company.

The time machine heuristic is a tool to make strategic business decisions. Strategic decisions are an organization's decisions on how to allocate its limited resources to achieve long-term goals, such as profit and market growth. Instead of devising some "x-year plan" that details every little goal and action and aims to utilize resources optimally, we argue that in a volatile, uncertain, complex, and ambiguous (VUCA) world, experienced managers and successful companies rely on a toolbox of heuristics to make strategic plans and decisions. These heuristics can be robust against the unpredictability of a fast-changing environment and effective in helping organizations achieve their goals quickly. Imitation is a key heuristic in the strategy toolbox.

Imitation as a Strategy to Speed up Economic Development

In the early days of the nation, the US was largely a backward, agrarian country with little manufacturing capacity. At the time, the most booming industry in the world was cotton and textiles, which had made Britain an economic superpower, and the most innovative technologies were water-powered mills and looming machines. Alexander Hamilton, one of the founding fathers of the US and the eponym of the famed Broadway show, decided to follow Britain's path of industrialization and lobbied Congress to pass the Patent Act of 1793, which encouraged knowledgeable foreigners to bring their skills and inventions to the country, for which they would be richly rewarded. Hamilton believed that this was the most efficient as well as realistic means by which the country could be industrialized because innovations cultivated domestically would be both too slow and too risky. Aspiring entrepreneurs from Britain responded quickly. One of them was Samuel Slater, who used to work in a textile factory in England. Slater brought his knowledge of advanced carding and spinning machines to the new continent, set up factories in Rhode Island, and became one of the richest Americans in the era (see figure 5.1). He is often credited as the father of the Industrial Revolution in the US.

The British government was furious about the loss of their technologies and skilled engineers and workers to the new continent. To counter it,

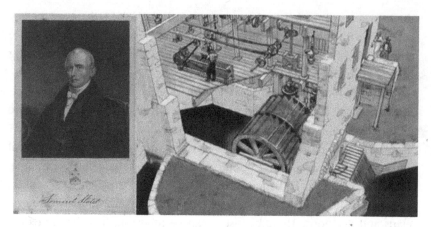

Figure 5.1
Samuel Slater (known as "Slater the Traitor" in Britain) and a diagram of Slater Mill, the first water-powered mill in the US. Source: https://www.nps.gov/blrv/learn/his toryculture/slatermill.htm.

they forbade local talent to leave the country and banned the export of all advanced machines, punishing people who did so severely. But one clever American, Francis Lowell, managed to bypass the tight controls by using his photographic memory ability to memorize design details of looming machines during his tours of English factories and reproduce these machines in the US with the help of a clockmaker. Working with business associates in Boston, Lowell built a new town in Massachusetts and made it the site of several factories. The town was later named after him, the great "stealer" of British textile technology.

The line between pirating and good-natured imitation is often murky. One may judge the acts of Slater, Lowell, and many individuals of their time as pirating and look at them with disdain. But how should one judge the behavior of the US government? Should it be condemned, or should one praise it for its audacity and strategic brilliance? In modern history, similar government-sanctioned pirating or mass copying has happened repeatedly. Britain itself had managed to get rid of its huge trade deficits with the Qing dynasty from the tea trade by smuggling tea plants from China to India and stealthily learning production techniques; Japan became a powerhouse in the electronics industry after World War II by buying a huge number of foreign patents and copying best-selling products in the Western market; and China gained its title as the "world's factory" partly by manufacturing products for foreign companies, learning the production techniques in the process, and by having domestic companies make and sell similar products.

Imitation is indispensable for the dissemination of an innovation, be it a water-powered mill, a vacuum machine, or a commercial platform. By imitation, a country or a company can reap the benefits of others' creations without having to take the risk of inventing something entirely new on its own. To be successful or even outcompete the original inventors, however, imitators paradoxically often need to be innovative in how they imitate.

Innovative Imitation

Theodore Levitt, who popularized the word *globalization*, wrote in a 1966 *Harvard Business Review* article titled "Innovative Imitation" that "no single company, regardless of its determination, energy, imagination, or resources, is big enough or solvent enough to do all the productive first things that will ever occur in its industry and to always beat its competitors to all the innovations emanating from the industry."[4] Therefore, companies should not be shamed or feel ashamed for imitating others. The smart thing is to accept the limits of their own innovative capability and strike a good balance between innovation and imitation.

For instance, in the early years of Google, its two founders, Sergey Brin and Larry Page, knew that they had the best search engine in the world, but they were reluctant to monetize it through advertising. Under pressure from their investors, however, they caved in. When they looked around for ways to generate revenue, they found that a rival search engine, GoTo.com, had made a lot of money by placing paid ads in prominent places in its search results and charging clients for actual clicks on the ads. Instead of coming up with a new model, Brin and Page decided to adopt this pay-per-click model. But they also improved it by adding some new twists and features of their own, such as giving a quality score to each advertisement to prevent spamming and switching to a more efficient bidding system for clients. The product was known as AdWords, and once it was introduced, the money started to roll in for Google.[5]

The tremendous success of the iPhone is another example of innovative imitation. Before the first iPhone was released in 2007 and became the best-selling smartphone, Blackberry was the undisputed pioneer and leader of the market. Blackberry phones could receive and send email, browse the

Internet, and allow other simple, web-based activities. These features were big breakthroughs from the old, not-so-smart mobile phones and a huge draw for business-oriented customers. Inspired by the success of Blackberry but unsatisfied with many design features (e.g., using a stylus as the input device) and the limited functionalities of Blackberry phones, Steve Jobs summoned a team at Apple to work secretly on a new smartphone that would be more powerful, user-friendly, and suitable for a broader range of consumers. Benefiting from the enormous technological reserves of Apple and the perfectionist attitude of Jobs, the iPhone was born with many innovative features and quickly drove Blackberry out of the market.

The iPhone, interestingly, then became the target of mass imitation, sprouting multiple companies that make similar-looking phones. Some of these companies, such as the Chinese brand Xiaomi, have managed to attract a horde of loyal followers of their own, making good profits in the competitive yet huge smartphone market.

Late Movers

Companies such as Xiaomi are late movers to a market pioneered by others. The pioneers enjoy certain advantages, such as preemptive positioning and strong brand association, but that does not mean that they are guaranteed to dominate the market forever. A study of fifty categories of products found that pioneers were more successful than late entrants in only about 30 percent of the categories.[6] For example, Jeep was the first car manufacturer to mass-produce a modern sport utility vehicle (SUV), a type of off-road, high-suspension vehicle. However, after the US Environmental Protection Agency relaxed its regulations on passenger cars in the 1990s, the demand for SUVs surged in the country. Almost all other car manufacturers, including foreign ones, quickly introduced their own SUV models. In 2021, the two models of Jeep, Grand Cherokee and Wrangler, were still selling well in the US market, but their sales were surpassed by SUVs made by Toyota, Ford, and other manufacturers.[7]

To be successful, it is not necessary for a company to be the most innovative and strive for the pioneer position. Imitating other companies and being a late entrant will also work, but one must be careful with the timing.

Late-mover heuristic: Monitor innovative products from other firms and imitate them as quickly as possible.

Levitt called it the *used apple policy*: "You don't have to get the first bite on the apple to make out. The second or third juicy bite is good enough. Just be careful not to get the tenth skimpy one."[8] With this policy, a firm watches the market response to a new product by another firm (the first bite). If it is bad (a rotten apple), there is no point in imitating it; if it is good (a juicy apple), then the firm should act quickly and get a profitable piece of it.

According to the strategy researchers Venkatesh Shankar and Gregory Carpenter, the late-mover heuristic should be especially useful (i.e., ecologically rational) for firms with relatively low financial resources and limited ability to innovate.[9] Japanese electronics companies in the 1960s and Chinese sportswear companies in the 1990s, for instance, fit this profile well, and some of them, such as Toshiba and Li-Ning, benefited tremendously from the late-mover heuristic. For firms with plentiful financial resources, some innovation on top of imitation works best. Apple's iPhone, Gillette's shaving razor, and Boeing's commercial aircraft are among the many examples of how resourceful and highly innovative companies can become market leaders despite moving late into their respective markets.

What to Imitate?

For imitation to work, it is important to select the right target. There are two major targets.

> *Imitate-the-successful: Imitate the best product, business model, or practice in a field.*
>
> *Imitate-the-majority: Imitate what most other firms do.*

The time machine heuristic and the late-mover heuristic are instances of imitate-the-successful.[10] They are helpful when it is clear what the successful targets are. When this information is not available or difficult to find, such as when no leader has yet emerged in a market, imitate-the-majority can be more helpful. In essence, the heuristic amounts to a company's bet that there must be good reasons behind other firms' decisions. When an American IT company wants to open its first call center overseas, for example, it must choose a country in which to site the center. With little experience but knowing that most companies with similar needs set up their call centers in India, the company quickly decides to do the same. Although following the majority does not always lead to the best decision, it can be of great help for companies that do not have the time or resources to conduct their own research. On many occasions, there will also be network externalities as more companies make the same strategic choice. In the case of call centers, as more companies choose India, a larger and more professional workforce is created there, which in turn attracts more companies and makes Indian call centers increasingly better.

As the saying goes, "When in Rome, do as the Romans do." A firm that wants to expand its business in a foreign market can also benefit from following the imitate-the-majority heuristic. Only in this case, the majority refers to the local customs, traditions, and businesses. For instance, two authors of this book lived in Singapore for many years. In a local KFC restaurant, one can find porridge, a traditional item on many Asian families' breakfast tables, on the menu. This might be strange for Americans who know and eat in KFC mainly for its chicken dishes. In an attempt to increase revenue and gain advantage over its main competitor, McDonald's, KFC first offered porridge in its China outlets in the 1990s. It became such a success there that KFC in other Asian countries, such as Singapore and Malaysia, gradually added porridge to their menus as well. More items inspired by Asian cuisine, such as Peking duck rolls and rice bowls, have since been added to KFC's menu (see figure 5.2). Interestingly, KFC restaurants back in the US have also started to provide these new creations to their American customers in recent years, to promote a healthier kind of food and change its greasy brand image. In the age of globalized business, stories like KFC's are happening everywhere.

Should Market Leaders Imitate?

Less successful firms can benefit from imitating the market leader. But should a leader imitate? It turns out that imitating a competitor's moves is also a strategy frequently used by leaders to defend their position in the market. In doing so, the leaders offload the burden of innovation and, importantly, reduce uncertainty about the market's responses to new products or technologies and consumers' ever-changing taste. For instance, Coca-Cola has been dominant in the Japanese soft drink market, but it nevertheless imitated Suntory, a close competitor. Specifically, instead of being the first to introduce a new product to the market, Coca-Cola often quickly imitated drinks initially introduced by Suntory (e.g., a seasonal cherry-blossom-flavored beverage shown in figure 5.3), fighting for whatever new market that the

Figure 5.2
KFC's "When in China, make what the Chinese eat" strategy. On the left is a promotion for the new "Peking Duck Spicy Roll" in China; the top-right photo shows a promotion for porridge in Singapore, and the lower right shows a promotion for three rice-bowl items in the US.

products are trying to gain. In another example, Intel was reported to have spent over $10 billion developing microchips for mobile devices. This was a move mimicking the strategy of its rising rival ARM, which had a niche in low-energy processors and had quickly encroached on the chip market with the rising popularity of smartphones, tablets, and autonomous vehicles.

Coca-Cola and Intel used the *second-mover heuristic* to help them maintain leading positions. Curiously, they chose different types of competitors to imitate. In the case of Coca-Cola, Suntory was similar to Coca-Cola, in that both had a broad product line and aimed to cover most of the market needs. Another Japanese soft drink company, Otsuka, which had roughly the same market share as Suntory, had a much narrower set of products, and its products were rarely imitated by Coca-Cola. In the case of Intel,

Figure 5.3
Coca-Cola imitating a rival's new product in the Japanese market. The left image is a
pink advertisement for a cherry-blossom-flavored beverage released by Pepsi, whose
brand license is owned by Suntory in Japan. The right one shows a pink advertisement
for a similar beverage released later by Coca-Cola. In both, the beverage is advertised
as seasonal and available for only a limited time.

AMD was the chip company most similar to it, as the two had been fighting
in the high-performance-chip market for years. ARM was the competitor
gaining ground in a market previously not covered by Intel, but Intel nonetheless decided to imitate ARM instead of AMD. Why?

The strategy researchers Dmitry Sharapov and Jan-Michael Ross addressed
this question by looking at simulated and real-world competitions.[11] They
found that it would be ecologically rational for leading firms to imitate their
closest challengers—that is, firms that are in second place and rising—when
environmental changes are frequent and substantial, such as in the fast-
developing market for computer chips. The main reason is that leading firms
can learn from the challengers where the market is heading, which would
be difficult to figure out otherwise, and use that to guide their next move. If
environmental changes are infrequent and minor, such as in the market for
soft drinks in Japan, leading firms should instead imitate competing firms
that share similar attributes with them. This is not only easier for the leading
firms to do but also helps them nip opportunities in the bud for the competitors that are most likely to threaten them.

The Adaptive Toolbox of Strategy Heuristics

Besides imitation, successful firms typically have a rich portfolio of strategy
heuristics and apply them adaptively, depending on their organizational

goals, the specific task, and the conditions of the business environment. Let's look at some of the heuristics that organizations use to make strategic decisions about acquisition, production, pricing, location, and market expansion.

Acquisition Strategy

A common strategy for large companies to remain competitive is to acquire other companies. This is especially so for IT companies because innovations appear frequently in the industry, and, as Levitt said, no single company can beat its competitors by always being first. When Cisco was taking off fast at the turn of the century, it used a satisficing rule to decide whether it should consider acquiring a company.

75–75 percent heuristic: Consider only companies that have no more than 75 employees, of which at least 75 percent are engineers.

Companies best suited to this rule were mostly start-ups that were highly innovative and likely already backed by venture capital. The rule worked well at the beginning, as Cisco was cash rich and could afford many investments that may or may not have worked. However, with the bursting of the Internet bubble and the company's long-term objectives more clarified, small companies with a large proportion of engineers were no longer essential. At this point, the 75–75 percent heuristic was replaced by a tallying rule based on five questions:

- Does the target share Cisco's vision of the industry's future direction?
- Does it have potential for short-term wins with current products?
- Does it have potential for long-term wins with future products?
- Is it located close to Cisco?
- Is its culture compatible with Cisco's?

The tallying rule was that an acquisition was given a green light if the target scored 5 points on these cues, a yellow light if it got 4, and a red light otherwise. It helped Cisco stay more focused by acquiring only companies that were a good fit. Cisco later modified the rule by dropping the "location proximity" cue. This simplification allowed Cisco to seize more opportunities in more distant markets.[12]

Production Strategy

When manufacturing capacity is limited and outside competition is fierce, firms must be strategic in prioritizing their production. That was exactly the situation that Intel faced in the 1980s, when Asian companies made aggressive moves in the computer chip market. Instead of basing its production decisions on complex optimization models, Intel relied on a one-clever-cue heuristic.[13]

Gross-margin heuristic: Manufacture products solely in order of their gross margins.

With this heuristic, Intel would avoid spending too much of their resources on products that were not profitable and whose market had become stagnant, such as the core computer memory that used to form the company's central business. The resources would instead go to highly profitable products that were in hot demand, such as microprocessors.[14] This heuristic is ecologically rational in a volatile market where product prices change wildly and focusing on the best-selling products is key to a company's survival. In a stable market where profit is reliable and competitors are fewer, the heuristic may hurt a company's long-term development by not having a more balanced product line.

The entertainment industry also needs to choose carefully what shows to produce, with many interesting ideas and scripts constantly floating around and begging to be made. Miramax, the movie company, relied on a fast-and-frugal tree to decide whether to green-light the production of movies in the 1990s.[15] Specifically, a proposal was rejected if the answer to any of the four questions in figure 5.4 was negative. With this production heuristic, movie proposals could be judged quickly on whether they would likely attract a broad audience. Almost all the movies that Miramax produced following this rule were box office hits, including such blockbusters as *The English Patient* and *The Talented Mr. Ripley*.

Pricing Strategy

According to standard economic theory, to maximize profits, firms should constantly update the prices of their products in response to changes in demand and supply. Although this practice is common in some companies such as airlines, others instead use simple rules that induce price stickiness. In the study of used car dealers in Germany mentioned briefly in chapter 3, the researchers found two versions of the satisficing heuristic in pricing:

Satisficing without aspiration-level adaptation: Set an acceptable price a for a car and sell the car to the first customer willing to pay this price or higher.

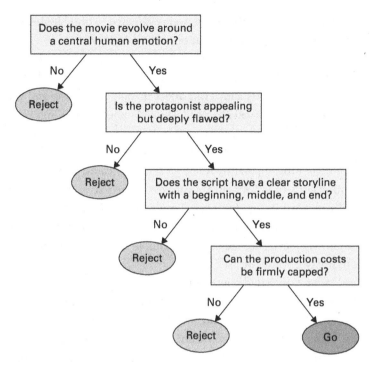

Figure 5.4
A fast-and-frugal tree capturing how Miramax decided whether to green-light the production of a movie or reject it. Note that this fast-and-frugal tree can alternatively be framed as a very strict tallying rule, in which the criterion for a "go" decision is that the answers to all four questions must be positive.

Satisficing with aspiration-level adaptation: Set the initial asking price at α. If the car is not sold in a period of β, then lower the price by γ.

Of the 628 dealers covered in the study, a whopping 97 percent used one version of the heuristic or the other.[16] The most frequent strategy was to set an initial price at the average price of similar cars and lower the price by 2 percent to 3 percent if the car was not sold within about four weeks. That said, individual dealers also adopted different β and γ values according to where their shops were located. For instance, dealers decreased the waiting duration β by about 3 percent for each additional competitor in the area and increased it by about 1 percent for each increment of €1,000 in gross domestic product per capita in the region. In general, the satisficing heuristics, with or without aspiration-level adaptation, were estimated to bring higher profits for the dealers than the best optimization-driven strategy.

Location Strategy

Where a business places its shops can affect its operation costs and profits greatly. Consider the fast-food chains McDonald's and Burger King. Between the two rivals, McDonald's had been the dominating "big brother" over Burger King, which led the two to apply different strategies in choosing store locations. Burger King apparently followed this one-reason heuristic: Always avoid locating close to McDonald's, regardless of the size of the market area. McDonald's, meanwhile, used this one: Locate close to Burger King in a small market area but far from it in a large area. According to a study by the marketing researcher Raphael Thomadsen, these simple heuristics brought mutual benefits to the two competitors.[17]

Beyond large, franchised companies, small business owners and executives must also make location choices from time to time. Forty-nine entrepreneurs in the Dallas area were interviewed about how they made location decisions.[18] The stakes were significant for each of them, and yet none claimed that they followed a process of searching and comparing alternatives extensively. Instead, 82 percent considered at most three locations and chose the best of those. Moreover, some entrepreneurs of small business projects used an imitate-the-majority heuristic by choosing to locate in an area where other firms in their industry had already located. The performance of these entrepreneurs was better than that of others who did not imitate.

Market Expansion Strategy

Companies have developed some powerful—some may say "sinister"— strategies to expand their business to new markets and new customers. One such strategy is the following.

> *Baiting heuristic: Provide a service or a product for free with a limited quality, quantity, or time, entice curious customers to try it, and then charge them for an upgraded service or product.*

The strategy has worked so well that it is now the go-to strategy for many subscription-based products, such as newspapers (e.g., the *New York Times*), cloud services (e.g., Dropbox), and e-commerce (e.g., Amazon). In some cases, customers can enjoy the "free" products permanently, so long as they agree, knowingly or unknowingly, that their data can be used by the product provider for any purposes that it wishes. Facebook and Google, for example, have been using this "pay with your data" business model to build an enormous base of customers, hoarding an astronomically large amount of data, monetizing it through advertising, and selling it to third parties.[19]

After its initial public offering in 2004, Google was eventually infused with so much money that it tried to dip its toes in many waters and drastically expand its business. Extending its success in search services, Google relied on the baiting heuristic to drive its expansion: start by providing a service for free or nearly free in a new field (e.g., 1 gigabyte of free storage for a Gmail account), attract a horde of customers to become a dominant figure in the field, and then find ways to make money. With this strategy, Google evolved into a multibusiness giant, and its current business includes email, online video hosting, browser, smartphones, laptops, autonomous driving, biotechnology, wearable devices, and artificial intelligence. The baiting heuristic is ecologically rational when the marginal cost for each new customer is negligible while the expected revenue per customer is constant.

The baiting heuristic is a case of what the organizational scholars Christopher Bingham and Kathleen Eisenhardt call a procedural heuristic. They interviewed executives of six companies in IT on their international market expansion strategies.[20] The executives' responses revealed a rich set of heuristics that could be grouped into four general types:

- *Selection* heuristics that guide what market opportunities to pursue (e.g., restrict internationalization to Asian countries or target only pharmaceutical companies)
- *Procedural* heuristics that specify actions on an opportunity (e.g., use acquisitions to enter new markets or leverage use of standards bodies when entering new countries)
- *Priority* heuristics that rank opportunities (e.g., prioritize government accounts or put more emphasis on the US market than others)
- *Temporal* heuristics that are related to the timing of opportunities or actions such as sequence, pace, and rhythm (e.g., market first in the US, then Japan, and then China)

How did the executives learn these heuristics? Apparently, they went through an elaboration-to-simplification cycle. At the beginning of the market expansion, the executives had only a few heuristics based on their previous experience. Having gained more experience in the new market, they then developed a large number of heuristics that were quite elaborate. With even more experience, however, the executives purposefully simplified both the number and the details of the heuristics. This pruning led to a repertoire of high-quality heuristics over time. Among the four types of heuristics, the executives learned selection and procedural heuristics faster than priority and temporal heuristics because the latter two required better understanding of the relationships among multiple opportunities and were cognitively more challenging. These findings suggest that these heuristics, as well as other heuristics covered so far, are the results of careful thinking and continued refinement. Although simple, they capture the practical wisdom executives have learned over years of experience, a topic that we examine further in chapter 13.

Smart Heuristics as Winning Strategies

Before its popularization in business, *strategy* was (and still is) a term used in wars and battles. Similar to what we have argued here, none of the great military strategy books, from Sun Tzu's *The Art of War* to Carl von Clausewitz's *On War*, have recommended just one grand winning strategy. Instead, they provide a repertoire of strategies and instruct what strategy to use in given situations, embodying principles of the adaptive toolbox and ecological rationality. The inherent uncertainty in almost all aspects of war is the main reason why. Although not as bloody as wars, business competitions are just as uncertain, especially in this day and age. Smart heuristics are winning strategies that can help firms survive and thrive in an increasingly competitive business world.

6 Innovation

Turning ideas and inventions into products and services that are successful in the marketplace—that is, innovating—is challenging. To accomplish this feat, innovative organizations rely on smart heuristics.[1] In fact, the term *heuristic* is etymologically related to creativity: in ancient Greek, εὑρίσκω (*heurískō*) means "I find, discover." The entrepreneurship researcher Mathew Manimala identified more than 100 heuristics used in entrepreneurial ventures.[2] He found that highly innovative ventures tend to use a different set of heuristics than less-pioneering ventures. For instance, compared to those of less-innovative start-ups, the heuristics of the innovative ventures are oriented more toward capability building, organic and integrated growth, people and value orientation, and continued exploration and learning. In this chapter, we examine a number of smart heuristics that help organizations become and remain innovative. Before we do so, though, let's first look at the conditions under which innovations occur in the first place.

Why Do Big Innovations Often Come from Small Start-ups?

One might expect that big innovations come mostly from big corporations with deep financial resources, large and well-trained staff, and excellent research and development (R&D) departments. But quite often, ground-breaking innovations come from small start-ups. Take Facebook: although it did not invent social media, it did revolutionize social interactions and our conception of friendship through the use of the Internet as a social medium at a global scale.[3] Google, meanwhile, transformed how we search information, PayPal revolutionized online payment, and Netflix changed the way we entertain ourselves with on-demand, online movies and shows.

None of these innovations were developed by big corporations. Postal services, for instance, have existed for over 2,000 years: Cursus Publicus, a state-run courier service, was established by the Roman emperor Augustus, and many European countries established postal services from the sixteenth century onward. One of their key original purposes was to enable the aristocracy and military to stay connected and share messages across great distances. Yet the public and private postal administration services that emerged after the nineteenth century and grew into the huge agencies that we know today have not been at the forefront of using the Internet for innovative communication.

Internet search engines did not come out of big media corporations such as NewsCorp or CNN. Big banks were not at the forefront of online payment. And online movie and television streaming via the Internet was not invented by Blockbuster, which had been the dominant brick-and-mortar DVD rental company. Reed Hastings, one of Netflix's cofounders, in fact was motivated to start Netflix after being frustrated about receiving a $40 late fee from Blockbuster.[4] These groundbreaking innovations, and many more, were driven by small ventures.

Uncertainty Is Necessary for Innovation

Why do big innovations tend not to come out of big corporations? We see at least three reasons. First, large organizations tend to have negative error cultures, an issue that we examine in more depth in chapter 11. For managers, this means that it is more important to avoid mistakes than to create innovative solutions. Avoiding mistakes can be achieved by "playing it safe": not taking risks, not trying new things, not threatening established products and services, and not being creative. Creativity by its very nature implies errors, as most new ideas and innovations fail, only some prove useful and commercially viable, and even fewer become blockbusters.

Second, and related, large organizations tend to encourage defensive decision making. A decision is defensive if managers pursue a second-best option rather than the best one, in order to protect themselves in case something goes wrong. To protect themselves, they gather long reports and unnecessary data, hire consulting firms, and generate stacks of documentation and paperwork before making a decision. Such a defensive culture stifles creativity and innovation.

Third, and most fundamentally, recall the distinction between risk and uncertainty from chapter 2. In a small world of risk, all possible actions, future states of the world, and their consequences are already known, so no innovation can happen. The economist Joseph Schumpeter popularized the idea of *creative destruction*, drawing on earlier work by Werner Gombart and Karl Marx.[5] Schumpeter argued that the gale of creative destruction continuously leads to innovation from within the economic system, incessantly destroying the old and creating the new. Underlying this process is inherent uncertainty, which limits organizations' ability to predict and control their environment, and thus their survival. Sooner or later—and sooner for most—organizations disappear, no longer being relevant or competitive.

That important insight is overlooked by theories that assume small worlds. It is also overlooked by large organizations that use their tremendous resources in an attempt to control every detail of their environment and manage risks. How can organizations thrive in this Schumpeterian world of creative destruction? Perhaps rather than expanding their risk management departments, companies could profit from establishing "uncertainty exploitation departments."

Under Uncertainty, Remaining Innovative Is Difficult

Some of the innovative start-ups mentioned earlier have since become big corporations and face the challenge of how to remain innovative. Many of them are unable to do so. Instead, they buy innovation by acquiring start-ups—the kind of organization that they were before they expanded and lost their edge. A German smart kitchen designer shared with us what happened after his small, cutting-edge company was awarded first prize in an international competition in California: he was approached by a manager from Google, who told him: "We will buy you or we will destroy you."

Buying innovation is challenging, however, because it is difficult to predict which new idea will end up being profitable. In 1998, when Netflix started, Blockbuster dominated the video rental industry in the US. In 2000, Netflix offered to sell itself to Blockbuster for $50 million. Blockbuster executives declined, as Netflix was losing money at the time, whereas Blockbuster was still profitable and flourishing. John Antioco, then the CEO of Blockbuster, opined that "the dot-com hysteria is completely overblown."[6] Meanwhile, Amazon CEO Jeff Bezos had offered to purchase Netflix earlier for around $15 million, but the Netflix owners declined the offer. That was the right decision: by 2020, Netflix had revenues of over $25 billion and a market capitalization of around $200 billion. Blockbuster, on the other hand, had to file for bankruptcy protection in 2010, and by 2014, all but one store, in Bend, Oregon, had closed.

Similar prediction failures abound under uncertainty. In 1876, Western Union, then the largest American telegraph company, did not purchase Alexander Graham Bell's telephone patent for $100,000, for the following reason:[7] "Bell expects that the public will use his instrument without the aid of trained operators. Any telegraph engineer will at once see the fallacy of this plan. The public simply cannot be trusted to handle technical communications equipment." Although this and similar examples are often presented in books as quirky decision failures, we need to recognize that hindsight is 20/20 and uncertainty entails big errors. There is no script for how the future will unfold.

Rather than trying to buy innovation, organizations could try to remain innovative. But how?

Innovation Heuristics

One large organization that has managed to stay exceptionally innovative over decades is 3M. Given its outstanding success, its practices deserve a closer look. Founded over 100 years ago, in 1902, as the Minnesota Mining and Manufacturing Company, 3M has maintained a high level of successful innovation ever since. Its product line contains over 60,000 products, including global household brands such as ScotchTape, Post-It Notes, and Scotch-Brite. Its sales surpass $30 billion a year. Impressively, the company reached the milestone of 100,000 patents in 2014, and this number has been growing by about 3,000 patents each year since then.[8] How could 3M be so innovative over such a long period of time? Several heuristics that the company relies on as part of its corporate culture appear to play a key role.

Driving Creativity with the 15 Percent Rule
A key one-clever-cue heuristic at 3M is the 15 percent rule. Launched in 1948, this rule allows scientists and engineers to spend around 15 percent of their work time on trying to innovate anything they like. William McKnight, who rose from a bookkeeping position to lead the company for many years,

provided the following rationale: "Encourage experimental doodling. If you put fences around people, you get sheep. Give people the room they need."[9]

15 percent rule: Scientist and engineers can spend 15 percent of their work time on trying new things.

This rule appears to waste resources and time. After all, this is paid time that is not used productively for concrete existing projects. And most of this experimental time is spent on generating ideas that will never turn into commercially viable products. Also, why 15 percent? And why 15 percent across the board? Clearly, some employees are more creative and innovative than others. To maximize innovation, it would thus seem to make more sense to allocate more time to highly creative employees and less or none to less creative ones. The problem with this maximizing approach, however, is that 3M does not operate in a small world of calculable risk, but in one where it is impossible to predict who will have the next creative idea that leads to a blockbuster product. It may well be an employee who hits on a great idea for the first time.

Many of the companies' patents and products, including the ubiquitous Post-It Note, were invented during 15 percent time. Employees in the infection-prevention division used it to pursue wirelessly connected electronic stethoscopes. As a result, 3M could introduce the first electronic stethoscope with Bluetooth technology that allows doctors to listen to patients' heart and lung sounds as they go on rounds, seamlessly transferring the data to software programs for deeper analysis. The product became highly profitable.

The 15 percent rule has been imitated—another effective heuristic!—by several other highly innovative companies such as Hewlett-Packard and Google. Bill Hewlett of Hewlett-Packard went so far as to cite 3M as a corporate role model when asked about a company that he greatly admired: "3M! No doubt about it. You never know what they're going to come up with next. The beauty of it is that they probably don't know what they're going to come up with next either."[10] And at Google, Gmail and Google Earth were conceived during the company's even more generous 20 percent time.[11]

Staying Innovative with the 30/4 and the 6 Percent Rules
Recognizing the temptation to become complacent, 3M instituted another one-clever-cue heuristic to stay innovative.

30/4 rule: 30 percent of the company's profits must come from products introduced in the last four years.

This rule constantly challenges the company not to rest on its laurels. Rather than being content with the profits from its existing patents and current blockbuster products, employees continue to search for new and even better products. This rule is complemented by yet another one-clever-cue heuristic.

6 percent rule: Spend approximately 6 percent of sales on R&D.

The 3M company spends a far larger percentage of sales on R&D than a typical manufacturer, providing resources to achieve the 30/4 rule. This has resulted in not only new products but also the creation of new entire industries. David Powell, 3M's vice president of marketing, affirms R&D's importance: "Annual investment in R&D in good years—and bad—is a cornerstone of the company. The consistency in the bad years is particularly important." The rule provides a fast, frugal, and transparent way to determine the R&D budget. It also motivates researchers and product developers to develop innovations that sell because when sales go up, more money is available for R&D.

Turning Failure into Success

The 3M company also has a rich tradition of telling the stories of famous failures that subsequently created breakthrough products. This tradition supports a culture of staying innovative and risking failure for uncertain rewards.

Failure-to-success heuristic: Whenever something fails, rather than accept failure, think about how to turn it into success.

Consider how ScotchGuard was invented: Patsy Sherman, a 3M researcher, was conducting experiments on fluorochemical polymers when a lab assistant accidentally spilled some of the mixture on her tennis shoes. She tried water, alcohol, and soap but was unable to remove the spillage from her shoes. From there, she got the idea that this substance could perhaps act as a stain protection barrier for other textiles.[12] After much additional experimenting, the formula for ScotchGuard was discovered, and the rest is history. Another failure-to-success story is the weak adhesive that was developed unintentionally in a project trying to develop a strong glue; instead of thinking of it as a failure and putting it aside, the invention was turned into the hugely successful Post-It Notes. Another 3M failure story from its early days forms part of the company lore: the company's initial business venture was to mine corundum, a material that they planned to use to make grinding wheels. Instead, what they found was an inferior abrasive. After much experimentation came their first breakthrough product: Wetordry sandpaper.

Learning from failure is supported by a healthy error culture. At 3M, this is expressed through the philosophy of Richard Carlton, 3M's director of manufacturing and the author of its first testing manual: "You can't stumble if you're not in motion." Similarly, 3M CEO McKnight formulated the "McKnight Principles" as the backbone of 3M's corporate culture. Its crucial passage reads: "Mistakes will be made, but if the man is essentially right himself, I think the mistakes he makes are not so serious in the long run as the mistakes management makes if it is dictatorial and if it undertakes to tell men . . . exactly how they must do their job."[13]

Trial-and-Error

Thomas Edison is often credited with having invented the lightbulb in 1879. However, an earlier version had been created many years before by the English chemist Humphrey Davy. Davy also identified the key challenge for making the invention a commercial success: finding materials and production processes that result in a cheap, brightly burning, and long-lasting product. It was only Edison who found the right combination of materials and manufacturing to make the lightbulb viable. To do so, he relied on the trial-and-error heuristic.

Trial-and-error heuristic: Try the first option that comes to mind. If it fails, try the next one. Repeat until you succeed.

Edison instructed his R&D facility to stock every kind of raw material imaginable. Combining the vast availability of materials with his experience with various production processes, he could try various combinations of materials. This heuristic process culminated in the invention of the Edison lightbulb.

Trial-and-error is deeply engrained in 3M's culture. As Carlton wrote, "Every idea should have a chance to prove its worth, and this is true for two reasons: (1) if it is good, we want it; (2) if it is not good, we will have purchased peace of mind when we have proved it impractical."[14]

Serendipity, the art of finding something that one was not looking for, plays an important role in innovation. Earlier, we shared how 3M's ScotchGuard originated in a lab accident in which a researcher, after a liquid intended for other purposes was spilled on her shoes, discovered its water-repellent quality. The discovery of Gore-Tex, a material that has become synonymous with high-quality, all-weather clothing, also happened serendipitously.[15] Instead of slowly stretching heated rods of polytetrafluoroethylene, as would normally be done, the company's founder, Bob Gore, at one point applied a sudden, accelerating yank. This stretched the material about 800 percent and created a microporous structure that was both waterproof and breathable—and Gore-Tex was born. Although this was, strictly speaking, not the result of a trial-and-error heuristic, serendipitous discoveries often happen during trial-and-error processes.

When is trial-and-error ecologically rational? Two key conditions are that the search for solutions is guided by expertise and that sufficient resources are available. The availability of resources enables companies to try all the potential solutions that they come up with. Expertise helps direct the search for solutions along a promising direction. This is the same logic as for the fluency heuristic (chapter 2): the first option that comes to experts' minds can be the best option.

When the trial-and-error search is not guided by expertise, it is not smart. Randomly trying solutions to see if any of them work can require a large number of trials and lots of resources. This approach is, therefore, more suitable for organizations that have those necessary resources (e.g., customers, materials, programmers, scientists). Because of the huge number of customer interactions and transactions, running in the tens or hundreds of millions a day, Amazon and other tech giants such as Facebook can run large numbers of little experiments, trying different things. For example, they randomly present web pages differing, among other attributes, in layout, color, and fonts to thousands of users to see which version leads to more income-generating purchases or clicks on paid ads. In conditions where resources are scarce or trials take longer, the random (unguided) trial-and-error heuristic may take too long or become too expensive. One context in which random trial-and-error is not ideal is heuristic design.

Heuristics for Product Design

The design of products is crucial from both aesthetic and cost perspectives. On average, it is estimated that about 70 percent of a product's cost is determined by its design.[16] How can products be best designed, given the almost infinite number of possibilities? Here again, heuristics play an important role. There are heuristics for the outcome of design (what a product should look like and what qualities a designed product should have) and for the process of design (how to come up with new designs).

Few companies are as famous for their designs as Apple. What is less known is that Apple, among countless other companies, was heavily inspired by the design principles of Dieter Rams, longtime head of design at the German consumer products company Braun. Figure 6.1 shows how Apple applied the *imitate-the-successful heuristic* (chapter 5) to design the iPod, closely imitating Braun's T3 transistor.[17] Using this heuristic, organizations imitate the best product, practice, or business model.

According to the ten principles that Rams formulated, good design is[18]

- Innovative
- Useful
- Aesthetic
- Understandable
- Unobtrusive
- Honest
- Long-lasting
- Thorough down to the last detail
- Environmentally friendly
- With as little design as possible

Rams's principles can be used as a tallying heuristic: designers can tally how many of the principles their design ideas met, with more being better. And companies could set thresholds as to how many principles have to be met for a design to be acceptable.

Process heuristics can help designers come up with new ideas. Colleen Seifert and colleagues have catalogued seventy-seven design heuristics in a qualitative study of industrial and engineering designers working on a variety of consumer products.[19] These heuristics include a whole range of approaches such as the following:

- *Add to an existing product*: Add an existing item to the product's functions. Consider physical attachment, creating a system, or defining relationships to products.

1958   2001

Figure 6.1

The imitate-the-successful heuristic used for product design. Apple's iPod imitates the design of Braun's much-earlier T3 transistor. Left, T3 transistor, from 1958; right, iPod, from 2001. This also illustrates the enduring power of simple design heuristics. Source: https://es.bellroy.com/journal/heroes-of-design-dieter-rams.

- *Bend*: Form an angular or rounded curve by bending a continuous material to assign different functions to the bent surfaces (see figure 6.2 for an example).

- *Expose the interior*: Show the inner components of the product by removing the outer surface or making it transparent for user perception and understanding.

- *Stack*: Stack individual components or make the entire product stackable to save space, protect the inner components, or create visual effects.

- *Unify*: Cluster elements according to intuitive relationships such as similarity, dependence, proximity, to unify them for visual consistency.

In subsequent research, Seifert and her colleagues showed that students could be trained in the use of design heuristics. The training was quite straightforward, consisting of a simple demonstration of design heuristics. As we discuss in chapter 13, an important advantage of heuristics is that they can be learned and taught relatively easily. After this training, students' designs were more creative than those of students in a control condition. The researchers concluded that design heuristics "appeared to help the

Figure 6.2
Example of the "bend" design heuristic: The bent bookshelf takes on a second function as an artistic interior design object. Source: https://www.etsy.com/uk/listing/155 846308/spiral-bookshelf-medium.

participants 'jump' to a new problem space, resulting in more varied designs, and a greater frequency of designs judged as more creative."[20]

Innovation through the Wisdom of Crowds

The wisdom-of-crowds heuristic can enable accurate predictions of future events, such as which candidate will win an election or which product will succeed in the market. It does so by taking the average of a large number of independent estimates (see chapter 3 for more details). Creativity and innovation, however, are based on being different from the norm or the average. Interestingly, a combination of brainstorming and wisdom-of-crowds can be used to develop successful new products. It has two steps:

Step 1 (brainstorming): Solicit creative ideas from your customers and collate them.

Step 2 (wisdom-of-crowds): Let your customers vote on these ideas and produce the winners.

Step 1 takes advantage of the diversity of ideas among large numbers of people. Step 2 then takes advantage of the popular opinion of a large number of people. Together, they make a powerful combination. Consider LEGO's application of this heuristic.

Few brands in the toy business are as innovative and successful as LEGO.[21] Millions of small, plastic bricks and figures are sold and assembled every year, bringing much joy to a large base of fans around the world and huge profits to the company. In the early 2000s, however, LEGO faced serious problems. Its design team had created increasingly complex products that required more uniquely designed individual components, and production costs soared. Meanwhile, sales kept dropping, as the sophisticated designs apparently did not sufficiently appeal to consumers. From 2002 to 2003, sales went down 30 percent.

In that critical situation, LEGO Ideas came into play. LEGO Ideas is an online platform where fans can share and discuss their own designs and vote for the designs they like. As soon as a design has gathered more than 10,000 votes, LEGO formally reviews it and decides whether it should go into production. Once it does, the product is distributed in LEGO stores, and the designer is rewarded with 1 percent of the net sales revenue. This crowdsourcing strategy to generate ideas takes advantage of the wisdom of crowds of LEGO fans. It has led to many good design ideas and products, mobilized fans' interest in a product before its launch, and increased fans' already strong loyalty to the brand. Soon after the launch of LEGO Ideas, sales rose again. In 2015, LEGO was the number-one toy company in Europe and Asia. In 2016, it sold 75 billion bricks.

Facilitated by the Internet, innovation strategies that aim to acquire good ideas from large crowds have become more and more plausible and doable. Sometimes it does not even require as sophisticated a setup as LEGO Ideas. Amazon started by selling books online before adding CDs and DVDs to its shopping catalog. With the whole platform in place, Jeff Bezos contemplated how Amazon should expand further. He then emailed 1,000 randomly selected customers and asked them what they would like to see the company sell. Many responded. Bezos recalled that one person wrote: "windshield wiper blades, because I really need windshield wiper blades."[22] At that point, Bezos realized that Amazon could sell anything online. The company went on to offer electronics, toys, and many other product categories over time. Apart from listening to their customers, Amazon also planned their expansion slowly, going one step after another without moving too fast. These heuristic strategies played a big part in Amazon's enormous success.

To Innovate or to Imitate?

Stephen Stigler's *law of eponymy* states that no scientific discovery is named after its original discoverer.[23] For example, the Pythagorean theorem, which explains the relationship between the sides of a right triangle, had already been known before Pythagoras. Similarly, Fourier transforms had been used by Pierre-Simon Laplace before Joseph Fourier, Siméon Poisson published the Cauchy distribution twenty-nine years before Augustin-Louis Cauchy "discovered" it by coincidence, and Bayes's rule was not discovered by Thomas Bayes. To make his point, Stigler noted that even Stigler's law was not discovered by himself (he credited the sociologist Robert K. Merton).

Stigler's law also holds in business contexts. Quite a few products are associated with names that are not their originators. As we saw in this chapter, Edison did not invent the lightbulb. And Mattel's famous Barbie doll was modeled after a comic-strip character in the German tabloid *Bild* named Lilli. Just as Lilli was designed to satisfy the taste of *Bild*'s adult male readers, Barbie satisfied prevailing gender stereotypes. The second version of Barbie in 1992 could speak, though perhaps it would have been better if she had not, given that she said things such as "Math is hard. Let's go shopping."[24]

Chapter 5 highlighted the value of imitation as a key strategy heuristic that enables firms to develop products, enter markets, and remain competitive. Even a highly innovative company such as Google imitated when it developed AdWords to monetize its search engine through the display of advertisements. Often, the line between imitation and innovation is blurry. Most organizations both imitate and innovate, as no organization can stem all innovation by itself. And although imitation is generally much easier than innovation, few organizations that completely lack innovativeness survive for long. The point is that both imitation and innovation are crucial for progress, and organizations rely on smart heuristics to both imitate and innovate.

7 Negotiating in the Real World

In the movie *The Negotiator*, Samuel Jackson plays an experienced police hostage negotiator, Danny Roman, who is falsely accused of corruption.[1] Desperate to find out the truth and prove his innocence, Roman himself takes several hostages inside the Chicago Police Department building. In one scene, Farley, a less experienced hostage negotiator, is trying to talk Roman into surrendering. Clearly tense, Farley asks Roman what he wants. Roman asks if he can see a priest, which Farley rejects.

> "That's good, Farley" says Roman "You shouldn't let me see a priest because a priest is associated with death and you don't want me thinking about death in my state, now do you?" "No, no." "But you also told me 'no', Farley. You can't say 'no'. Never say 'no' in a hostage situation . . . it's in the manual . . . never use no, don't, won't, or can't . . . it eliminates options and the only option that leaves is to shoot someone."

Roman then proceeds to get Farley to say no a few more times before setting an ultimatum that he will shoot one of the hostages if Farley says "no" one more time.

The movie scene illustrates something important: Good hostage negotiators use heuristics. Farley acted in the way that many novices would: He tried to engage Roman in conversation to find out what he wanted and what his motives were. He tried to extract information from Roman to analyze the situation. However, he was not good at adhering to simple but critical rules, such as to never say "no" to a hostage taker. Unaware of such simple rules, unskilled hostage negotiators can easily fall into certain traps that escalate the situation. The use of heuristics is not limited to hostage negotiators; it applies to negotiations more broadly.

Textbook Folklore: Heuristics Make Negotiators Biased

Negotiation is a process through which two or more parties resolve conflicts by jointly agreeing on how to allocate scarce resources.[2] Negotiators use a toolbox of heuristics, such as meeting in the middle, making ambitious opening offers, and imitating the counterpart, as we show in this chapter. This should not come as a surprise, as negotiations are often characterized by conditions such as complexity, uncertainty, and time pressure that require heuristics.

Yet reading any standard negotiation textbook, you would get the impression that successful negotiations are all about analysis and that heuristics are the worst enemies of negotiators, worse even than emotions. Whereas negotiators can at least use emotions strategically to manipulate their counterparts—for example, by pretending to be angry—heuristics are virtually always linked to cognitive biases.[3] For instance, according to these textbooks, the availability heuristic leads negotiators to be overly influenced by readily available or easily retrieved information, such as the sticker price on a used car, at the expense of more critical but less salient information, such as the car's actual market value.

This negative view of heuristics is then followed by the advice that negotiators should avoid using heuristics themselves and instead exploit the "heuristic reasoning on the part of others for personal gain."[4] The problem with textbook advice on how to negotiate is that it is based almost entirely on two kinds of "small-world evidence":[5]

- *Complex theoretical models of abstracted small-world problems based on game theory or negotiation analysis:* these provide "rational" solutions but not practical advice for how negotiators should behave to achieve these solutions.

- *Laboratory studies of small-world negotiation simulations mostly with non-expert negotiators:* these provide empirical insights into the factors that influence the outcomes of hypothetical or small-incentive negotiation simulations but offer no guarantee that these insights generalize to actual, large-world negotiations outside the lab.

Both approaches, one theoretical and one empirical, are applied in small worlds where the state space is given. In a typical lab study, negotiators receive clear instructions on which aspects of the deal to negotiate (e.g., price, quality, or delivery date), which agreements are possible, and how many points

Table 7.1
A small-world negotiation illustrated by a typical negotiation point sheet used in laboratory negotiation research

Issue	Option	Recruiter Points	Candidate Points
Salary	$80,000	−4,000	4,000
	$75,000	−2,000	2,000
	$70,000	0	'0
	$65,000	2,000	−2,000
	$60,000	4,000	−4,000
Location	New York	800	800
	Boston	600	600
	San Francisco	400	400
	Houston	200	200
	Chicago	0	0
Moving expense coverage	100%	−600	1,200
	90%	−450	900
	80%	−300	600
	70%	−150	300
	60%	0	0
. . .	. . .	. . .	. . .
	. . .	. . .	. . .

In this negotiation simulation, a recruiter and a candidate negotiate a job offer. Note that each party knows (a) all the negotiation issues (eight in total in this simulation, only three of which are shown in the table); (b) all the possible agreement options; and (c) the points they receive for reaching an agreement on each issue. This situation differs from most real-world negotiations, where considerable uncertainty exists regarding the issues, options, and payoffs.

they get for each agreement (see table 7.1 for an example). They do not know the instructions and payoff scheme of their counterpart (and vice versa). In theoretical analyses, the modeler knows, or even sets, the payoff schemes for both parties and then proceeds to determine optimal solutions. In such situations, one can know what is best, and negotiation analysis as a normative endeavor can determine optimal agreements.

However, real-life negotiations are conducted under uncertainty and complexity. Consider United Nations negotiations on climate action that involve negotiation teams from more than 100 countries.[6] The teams need to reach a consensus to address climate change, a complex problem with environmental as well as political, social, economic, and health-related dimensions. This is not the kind of "small world" to which optimization applies. Even for less-complex business negotiations, uncertainty abounds. Multiple issues need to be agreed on, such as price, quantity, quality, warranty, and service conditions, but it is often not clear up-front what all the relevant issues are, and even less clear what the agreement space is, that is, what possible solutions exist. In addition, negotiating parties can change their mind midway as the situation changes (e.g., raw materials become much more expensive), making a previous offer untenable; or a competitor makes an offer, disrupting the negotiation process.

When negotiations are conducted under uncertainty, there is no way to determine an optimal solution. Negotiators have to rely on heuristics. The only question is: In what situations should they rely on which heuristic? That is the question of ecological rationality. As Reinhard Selten made clear (see chapter 1), the best solution in a small world is not the best in the real, uncertain world. That distinction is often forgotten in textbooks. Having failed to distinguish between small and large worlds, they provide a one-sided portrayal of negotiators as biased. This tendency to see biases everywhere, even when there are none, has been called the "bias bias."[7]

How Skilled Negotiators Plan and Act

In a refreshing exception, the negotiation researcher Neil Rackham studied the behavior of successful negotiators.[8] Here, success was not based on the number of points achieved according to predetermined payoff schemes in well-defined negotiation simulations. Instead, Rackham studied forty-eight skilled professional negotiators over 102 negotiation sessions. These included seventeen union representatives, twelve managers, and ten contract negotiators (along with nine others). To be included, the negotiators had to satisfy all of the following three criteria:

- Be rated as effective by both sides.
- Have a track record of success over a substantial period of time.
- Have a low incidence of implementation failure.

Negotiators who did not satisfy these criteria or whose data were not available were placed in the "average" group. Rackham analyzed behaviors during two stages: preparation and actual negotiation. In the preparation stage, results show that in comparison to the average negotiators, skilled negotiators

- Considered about twice as many possible outcomes and options (5.1 vs. 2.6).
- Spent about three times as much time on anticipated areas of agreement (37 percent vs. 11 percent).
- Considered about twice as many long-term issues (8.5 vs. 4.0).
- Planned specific sequences (e.g., negotiating issue A, then B, then C) about half as often (2.1 vs. 4.9).
- Set ranges ("We aim to get $2 but would settle for $1.80") instead of a fixed point more often.
- Spent a similar amount of time on preparation, suggesting that they planned not more often, but more efficiently.

These findings suggest that skilled negotiators paid more attention to the uncertain nature of negotiations. They planned for ways to increase flexibility by thinking about more options, setting ranges, and avoiding fixed sequences. Flexibility enhances negotiators' ability to maneuver under uncertainty when one does not know what is going to happen next. Skilled negotiators were also more concerned with long-term viability, as uncertainty tends to increase over time, and focused more on areas of agreement, given that uncertainty can lead to more misunderstanding and conflict.

When analyzing behaviors during face-to-face negotiations, Rackham found a continuing theme from the planning stage that skilled negotiators avoided creating disagreement and conflict. They did so by doing the following:

- Using only about one-fifth as many irritator words as the average negotiators (2.3 vs. 10.8). Irritators are words such as "generous offer," "fair," or "reasonable," which refer to one's own offers or behaviors. Such words can irritate a counterpart, who takes the words as implying that they are unfair, unreasonable, or not generous.
- Making only about half as many immediate counterproposals (1.7 vs. 3.1). Immediate counterproposals tend to be seen by the counterpart less as proposals than as attempts to block or oppose their proposals at a time when they would like to discuss them.
- Using verbal attacks or defenses only about one-third as often (1.9 vs. 6.3). Verbal attacks or defenses can lead to spiraling behaviors, as when

one party does something the other party sees as an attack so they defend themselves, which the first party sees as an attack, and so on, leading to escalating conflict.

• Using behavior labeling in areas of disagreement only about one-third as often (0.4 vs. 1.5). Whereas an average negotiator might say, "I disagree with you about . . ." skilled negotiators are more likely to begin with the reasons and then state their conclusion without labeling it as a disagreement, even when they are in fact disagreeing.

Skilled negotiators also engaged in activities that increase clarity and understanding and reduce uncertainty about their intention and other internal states, such as the following:

• Applying behavior labeling in areas other than disagreement more than five times as often as average negotiators (6.4 vs. 1.2), by saying phrases such as "Can I ask you something" and "If I could make a suggestion." Note that average negotiators label disagreement more often (1.5) than skilled negotiators (0.4). Therefore, behavior labeling seems ecologically rational in most cases, but not when disagreeing, when more uncertainty may be better.

• Making statements about their internal states, such as feelings, doubts, and motives, about 50 percent more often (12.1 vs. 7.8). In doing so, they made the invisible visible, thus reducing uncertainty and increasing trust.

In addition to behaviors that reduce uncertainty about themselves, skilled negotiators engaged in behaviors that reduce uncertainty about the other party by doing the following:

• Asking questions more than twice as often as the average negotiators (21.3 vs. 9.6). Asking questions is a key strategy to gain information and reduce uncertainty about the other party's interests, goals, and views. It can also help uncover creative options for agreements.

• Testing their understanding twice as often (9.7 vs. 4.1) by making inquiries such as "Do I understand you correctly that . . ." These questions not only reduce the chance of misunderstanding but also are subtle ways to question a statement or position of the counterpart, prompting them to reconsider and restate it.

• Summarizing previous points about twice as often (7.5 vs. 4.1).

Skilled negotiators were also more concerned about being able to implement the agreement, and they wanted to be sure that an agreement was not based on a misunderstanding. In contrast, average negotiators could have been so anxious to reach an agreement that they might have preferred not to endanger it by uncovering latent disagreements.

Finally, skilled negotiators appeared to understand better that less can be more. Specifically, they focused more on quality over quantity when providing reasons to support their argument or proposal:

- Skilled negotiators used fewer arguments to back their case (1.8 vs. 3.0).

In focusing on their strongest reasons, skilled negotiators can avoid what is called *argument dilution*: adding a weaker argument to a stronger argument increases the total number of arguments but dilutes the overall potency of the argumentation. Moreover, given that negotiations are strategic interactions, adding a weak argument provides counterparts with the opportunity to pick and pounce on the weak argument while shifting attention away from the strong argument.

Rackham's research suggests that skilled negotiators are well aware that they are acting under uncertainty. As we know, heuristics are effective decision-making strategies under such large-world conditions. Next, we discuss some of the most common negotiation heuristics.

Heuristics for Negotiating Successfully

In an intriguing twist, the very same textbooks that warn their readers of the dangers of "heuristics and biases" often promote the use of heuristics, albeit under a different name: *negotiation strategies*. For example, logrolling is an integrative ("win-win") negotiation strategy in which negotiators use the heuristic of giving the other party what it cares more about and they themselves care less about, so that both parties are better off. Warning negotiators not to use heuristics while at the same time recommending heuristic strategies is not only self-contradictory but also not fruitful. A more valuable path would be to recognize negotiation strategies as heuristics and analyze their ecological rationality. We begin with a heuristic that we encountered in chapter 5, where it was used for a different purpose, pricing.

Satisficing

In distributive ("win-lose") negotiations, the bargaining zone is fixed, and what one party wins, the other party loses. Thus, it is purely a matter of distributing a fixed bargaining zone. A commonly recommended strategy is to determine one's desired price (the target price) and a reservation price—that is, the highest acceptable price for a buyer and the lowest acceptable price for a seller—and hold onto both throughout the negotiation. Setting a target price is not the same as a maximization strategy, where more is always better and there is no "good enough." Maintaining the reservation price inflexibly, meanwhile, violates the idea of Bayesian updating, by which negotiators should revise their reservation price based on information received during the negotiation.

In fact, this strategy is a version of Simon's classic satisficing heuristic. The buyer has an aspiration level of a desired buying price, the target, which is not revealed. The buyer starts with an offer below the target and ultimately accepts any offer that is better than the aspiration level. If, after some exchange, the buyer cannot get their target, they may then revise offers, but not above their reservation price.

The target–reservation heuristic (from a buyer's perspective) is as follows:
Step 1: Set an aspiration level (target) α and a reservation level β ($\alpha < \beta$).
Step 2: Start negotiation with a price below α and make concessions successively.
Step 3: Agree at a price lower than β; otherwise, end negotiation.

When it is uncertain whether information can be trusted and it is impossible to know the counterpart's situation for sure, such simple rules can help achieve a satisfactory negotiation performance, while at the same time protecting against major risks and losses.

Imitation

Imitation can help negotiation in at least two ways. In the *imitate-the-successful heuristic*, novice negotiators imitate experienced, successful negotiators, similar to how organizations imitate the best firms and practices in their industry, as discussed in chapter 5. Second, imitation can also be an effective heuristic in negotiations itself. For instance, the negotiation researcher William Maddux and his colleagues found that negotiators who imitated their counterparts' mannerisms, such as their body movements, improved their negotiation outcomes.[9] In online negotiations, linguistic mimicry—that is, imitating the language of the counterpart—also improved outcomes.

Mirror heuristic: Imitate the mannerisms and language of your negotiation counterpart.

Mirroring appeared to be particularly effective at the beginning of the negotiation, not at the end, which illustrates an ecological boundary condition signaling the importance of first impressions.[10] This process is reminiscent of humans' fundamental ability to experience empathy, and it may be facilitated by mirror neurons.[11]

Reciprocity

Reciprocity is a specific form of imitation in which imitation is directed toward the originator of the action. This is expressed in the notion of "give-and-take": one party gives something, and the counterparty receives; the roles are then reversed and the behavior is imitated, such that the counterparty gives something.

Reciprocity heuristic: When a party engages in a behavior toward you, reciprocate the behavior toward them.

Positive reciprocity applies to situations in which we receive something we value, and negative reciprocity when others hurt us in some way and we retaliate. In both cases, a sense of equity is restored. Positive reciprocity in particular has been credited with creating social capital. Giving and taking equally over time forms the basis of social exchange and trust.[12] Imagine that two neighbors exchange gifts that are worth $20 each. At first glance, it appears like a zero-sum exchange, with each of them being as well off as they were before the exchange. However, at a deeper level, both gain something: they strengthen their bond and the confidence that if one helps the other, the other will return the favor. In this way, the neighbors build trust and create social capital. If they do so repeatedly and across different people, a higher level of trust within a community or society develops, making everyone better off. Many societies have developed specific rules governing how much to reciprocate under different circumstances. In gift-giving cultures such as Japan, reciprocity is constantly practiced and taught from a young age, making it a deeply engrained heuristic of social relations.[13]

Reciprocity also plays an important role in negotiations. Reciprocity manifests itself as the expected give-and-take. Even a tough negotiator such as J. Paul Getty, the founder of Getty Oil Co., appreciated this advice: "My father said, 'You must never try to make all the money that's in the deal. Let the other fellow make some money too, because if you have the reputation for always making all the money, you won't have many deals.'"[14]

Reciprocal concession making is an effective strategy for moving toward an agreement, step by step.[15] Imagine a seller asking $4,000 for a used piano and a buyer offering only $3,000. How are they to come to an agreement? Each party could insist on their price, which would likely result in the other party walking away, with no deal reached. They could meet in the middle right away, but there is a danger in that as well: if the buyer proposes $3,500, the seller might counteroffer with $3,750. To avoid such undesirable outcomes, negotiators often proceed in steps of relatively small reciprocal concessions. For example, the seller might counteroffer $3,800 and the buyer reoffer $3,200. Often, negotiators make their concessions smaller over time to indicate that they are reaching their reservation point, thus providing a clear signal to the counterpart.[16] This "negotiation dance" then continues until the parties meet. Although it may appear inefficient, it is remarkably effective in situations of uncertainty.

> *Reciprocal concession heuristic: Reciprocate the concession(s) of your counterpart until both parties meet at a mutually acceptable agreement.*

The reciprocal concession heuristic works well if both parties' starting offers are about the same distance away from the fair market value. However, if one party starts with an extreme first offer and the other starts with a moderate first offer, reciprocal concession making can end up in a deal that advantages the first party or leads nowhere, as the latter party may not be willing to reciprocate concessions.

The Dilemmas of Trust and Honesty

The dilemmas of trust and honesty further illustrate the power of the reciprocity heuristic combined with the satisficing heuristic. The dilemma of honesty refers to the decision of how honest and transparent to be with the counterpart. The dilemma of trust presents the flip side of the dilemma of honesty and refers to the decision of how much to trust the counterpart. It is a dilemma because if one trusts everything the counterpart says, one may be taken advantage of by a cunning negotiator; however, if one does not trust anything the counterpart says, it becomes almost impossible for the two sides to reach an agreement.

Under some circumstances, the dilemmas may not exist. For example, when negotiators have made numerous deals with each other in the past and have always lived up to their agreements, there is good reason to predict that they will do so again. But what if the interaction experience is limited, especially when the parties have never negotiated with each other before? A solution to the dilemmas is to adopt a form of stepwise reciprocity: revealing some sensitive information and trusting the counterpart a bit, and then observing if the other party behaves in a trustworthy manner. For example, one party could reveal that they are under time pressure to make a deal urgently. If the other party shares that they also prefer to come to a conclusion quickly—rather than trying to take advantage of the situation—it is a sign that they can be trusted. Over time, negotiators can take multiple such steps until both parties are satisfied that they can trust each other.

Well-Defined Games

So far, we have discussed negotiation heuristics that work well in large, real worlds. Most of negotiation research, however, takes place in small worlds, where all possible actions and their consequences are known for sure. This holds more generally in research on cooperation and conflict management in experimental games. How do heuristics fare in these situations? When one of us (Reb) was a PhD student at the University of Arizona, a number of experimental economists, including Amnon Rapoport and Vernon Smith (who won the Nobel Memorial Prize in Economics in 2002 for his seminal work on experimental economics), regularly ran experiments in the Economic Science Lab. As participants, we received detailed instructions, sometimes pages long, and could proceed only after passing a test of our understanding. In all cases, monetary incentives were tied to the outcomes of our decisions. The reason for this is that in game theory, it is important that decision makers are clear about the rules and incentives. Moreover, not only is everyone supposed to know everything (all possible options of all players and all their payoffs), but every player is also supposed to know that every other player knows everything. This is called *common knowledge*, which is meant to ensure a small world.

While admiring the researchers' efforts, as participants we often had other goals in mind in addition to earning money, chiefly getting out of the lab as quickly as possible and returning to our desks. After several experiments, we noticed that the payoffs among participants varied only slightly (because we knew each other, we were able to compare the results afterward). Moreover, there was relatively little control over payoffs, which depended not only on one's own choices but also on those of others. As a result, after some experience playing these games, a subjectively "rational" strategy for at least some participants was to choose very quickly to reduce time in the lab, given roughly the same expected earnings and small variations. These experiments typically found that participants did not behave rationally according to the predictions of rational choice theory. Ironically, we may have been more rational than the researchers gave us credit for, minimizing the amount of time spent to receive a reasonable payoff.

This example illustrates at least three things. First, in a small world, maximization—or, as in this case, minimization (of time spent)—is possible. Second, even in a small world, where the options are well defined, it is difficult, if not impossible, to control for what matters to people. We cared not only about the monetary payoffs but also about our time. This was not modeled in the analyses of the data. Finally, to reduce the amount of time spent, participants could use simple heuristics. These included choosing randomly, choosing cooperatively, and choosing the first available option. These heuristics are ecologically rational under conditions where payoffs depend little on choices and the relationship between the two is uncertain. Thus, despite a small-world setting that is in principle amenable to optimization, we still relied on simple heuristics. Depending on the decision goal, these heuristics can outperform highly complex strategies, even in small-world economic games.

Tit-for-Tat

Consider the famous prisoner's dilemma game (figure 7.1).[17] In the game, two persons who committed a crime together are interrogated separately. If both remain silent (cooperate), both receive light sentences, as the evidence is weaker. If both talk, they are both charged and have to stay in prison for longer. If one of them talks (defects) but not the other, the one who talks is released and the other receives a more severe punishment. In this situation, regardless of what the other party does, economic rationality suggests that one should always talk. However, this leads to the paradoxical and undesirable outcome that both parties are worse off than if both remain silent.

Player B

	Cooperate (silent)	Defect (talk)
Cooperate (silent)	(−1, −1)	(−3, 0)
Defect (talk)	(0, −3)	(−2, −2)

Player A

Figure 7.1
The tit-for-tat heuristic can outperform highly complex strategies. In the prisoner's dilemma game, two parties can either defect or cooperate. Each has an incentive to defect because it would lead to a better outcome, no matter what the other party does. However, if both defect, they receive a worse outcome (−2) than if they both cooperate (−1). In the iterated version of the game, the simple tit-for-tat heuristic, in which a player begins by cooperating and then imitates the decision of the counterpart from the previous round, performs remarkably well and has repeatedly beaten highly complex strategies.

In the *iterated* version of the game, in which two parties play the game repeatedly, there is no simple theoretical solution on what the best move would be. One of the great surprises from a number of computer simulation contests of this game was the performance of a simple heuristic.

Tit-for-tat: Cooperate first, and then imitate the opponent's move.

This heuristic recommends being kind first and then imitating the other side's actions. It is based on two principles: cooperation and imitation. Despite—or perhaps thanks to—its great simplicity, tit-for-tat won several computer tournaments after being pitted against various more complex strategies.[18] Therefore, even in the small world of the prisoner's dilemma, simple heuristics can perform extremely well and, moreover, be highly robust in the sense of working well against a whole array of other strategies.

Studies on the ecological rationality of tit-for-tat have revealed conditions under which it is less successful, such as when other parties make mistakes. In these situations, the heuristic can easily be adapted by using the more forgiving *tit-for-two-tats*. Alternatively, switching to another simple heuristic, *win–stay, lose–shift* (which repeats a choice if it met the aspiration level in the previous round but changes if not) can be useful as well.[19] For many classes of symmetrical two-player games, imitating the opponent's behavior if it was successful in the previous encounter is an unbeatable strategy.[20] Exceptions are games of the rock-paper-scissors variety, which again show a boundary condition of the ecological rationality of imitation.

1/N

Consider another small-world game known as the *ultimatum game* (figure 7.2). Here, one person, the proposer, gets a certain amount of money (say $10) from the experimenter. The proposer then is instructed to offer any amount between $0 and $10 to another person, the receiver. The receiver can either reject the offer or accept it. If accepted, each person gets their share; if rejected, neither player gets anything. Economic theory suggests two very clear predictions. First, the receiver should accept any amount larger than zero, as at least some money is better than nothing. Second, it follows that the proposer should or needs to offer only 1 cent or other minimum level of allowable amount and keep the remainder for themselves.

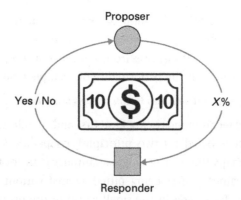

Figure 7.2
The 1/N and 1/N minus delta heuristics describe behavior in the ultimatum game. In this game, a proposer can offer a fraction of a given amount (e.g., $10) to the responder, and the responder can decide whether to accept or reject the offer. If the responder rejects, neither party receives anything. According to economic rationality, a responder should accept any offer above zero. The empirical finding is very different: Responders commonly reject offers of 1 cent and other highly unequal amounts. Likewise, proposers rarely offer widely unequal offers, with the most common being 50 percent (1/N) or slightly below 50 percent (1/N minus delta).

As scores of studies have shown, though, most people do not follow this selfish strategy.[21] Proposers do not typically offer only 1 cent. Likewise, responders commonly reject 1 cent and other small offers, deeming them unfair. Ask yourself: would you accept such a measly offer? When a friend of ours tried to sell a fairly expensive opera ticket literally at the last minute, a well-dressed man offered her a ridiculously low amount just before the doors closed. Our friend was so offended that she tore up the ticket in front of him.

To explain people's behavior in this game, highly complex versions of utility maximization have been proposed, such as the inequity aversion model.[22] However, these models are deliberately as-if, just like utility maximization models in general. That is, they are not intended to describe the actual process of decision making in the ultimatum game (or by our friend). They also have been criticized for not being able to predict participants' choices but only to fit the parameters to the data, after the fact.[23]

Empirically, the most common offer is at or slightly below 50 percent, or $5 in our example, and this amount is almost always accepted by the receiver. Thus, players' decisions can be parsimoniously explained by an equality heuristic: As the proposer, offer a fair amount to the counterpart, and as the responder, reject any unfair amounts. This amount can be determined by the $1/N$ heuristic: split the total amount ($10) equally across all parties (two parties, thus $5 each). Alternatively, the proposer can use a $1/N$ minus delta variant: first split the amount equally, and then subtract a small amount (delta) from the offer.

Ecological Rationality in Negotiations

The movie *The Negotiator*, from our opening example, also features Kevin Spacey in the role of a senior police hostage negotiator named Chris Sabian, the counterpart to Danny Roman. The scene introducing Sabian shows him unsuccessfully trying to resolve a conflict between his wife and their young daughter. Frustrated, he mumbles: "You know, I once talked a man out of blowing up the Sears Tower, but I cannot talk my wife out of a bedroom or my kid off the phone." His wife sarcastically retorts: "That's because no one is standing behind you with a big gun."

The scene illustrates the contextual nature of negotiation skills: Despite his excellent track record as a top hostage negotiator for the New York Police Department, Sabian's negotiation strategies are of little use with his wife and his young daughter—the situation is entirely different. Although we have seen that negotiation heuristics do well in both large and small worlds, they are not universal tools; they are ecologically rational only in certain situations.

Take equality heuristics. Dividing a fixed pie into equal pieces ($1/N$) and meeting-in-the-middle are key heuristics in negotiation. A comprehensive review shows that sticking to the principle of equality can enhance perceptions of distributive justice during negotiation, which not only facilitates successful outcomes but also helps nurture enduring relationships between the negotiating parties.[24] However, in win-win integrative, as compared to fixed-sum distributive negotiations, using an equality heuristic such as meeting-in-the-middle may leave money on the table. Here, negotiators could fare better by trading off on their differences using integrative heuristics such as logrolling.[25] Thus, using an equality heuristic is more ecologically rational in distributive negotiation situations, whereas using logrolling is more ecologically rational in integrative negotiations.

Consider another situation: Who should make the first offer in a negotiation? Whereas some sources recommend never making the first offer, others advise the opposite. Both recommendations can be bolstered by plausible arguments. For example, making the first offer can be advantageous by setting an anchor.[26] But letting the other party make the first offer can be used to sound them out, despite the possibility that the other party will make a low anchoring offer. How is a negotiator to decide what to do? It depends on the situation: Making the first offer is more likely to backfire when it reveals information about compatible preferences. In such situations, negotiators may do better by letting the other side make the first offer.[27] First offers can also backfire when they are perceived as too extreme. For instance, the negotiation researcher Martin Schweinsberger and colleagues found that counterparts felt offended by extremely low opening offers. However, low-powered counterparts were more likely to walk away from the negotiation, resulting in an impasse, whereas high-powered negotiators were more likely to continue to negotiate. Thus, making extreme first offers may be more advisable in situations with powerful counterparts but not with powerless ones.[28]

Negotiating in the Real World

If you have ever visited a street market in many parts of the world, you will know that bargaining is not just possible, but expected. In fact, it would be impolite *not* to negotiate. Not negotiating may signal that one does not find the other person worth engaging with. One of us (Gigerenzer) once visited a glass-blowing shop in Murano in the Venetian Lagoon together with a friend, who fell in love with two spectacularly beautiful and expensive lampshades. At her request, he began to negotiate with the shop assistant, praising the products but offering half the price. After some five minutes of offers and counteroffers, the shop assistant, acknowledging Gigerenzer's tough bargaining skills, handed him over to the shop owner. Over tea and exchanges of compliments on the prospective buyer's taste and the beauty of the lampshades, the gap between the asking price and the offer narrowed until it no longer budged. At this point, both sides had experienced so much pleasure in the process of bargaining that neither wanted the sale to fall through. Finally, the owner offered a wager: to toss a coin, and if Gigerenzer won, he would get the lampshades at the price he offered, and if the owner won, the owner would get his asking price. To the great excitement of the shop assistants, who appeared to have never seen their boss make such an offer, he tossed the coin. It came out in favor of the owner. Independent of the outcome, both sides enjoyed the social encounter of negotiating more than the result.

Negotiation courses in business schools often abstract from the richness of such real-world negotiations to employ elegant theoretical concepts such as pareto-optimality. In contrast, skilled negotiators flexibly draw on their adaptive toolbox of negotiation heuristics, depending on the negotiation situation (e.g., distributive versus integrative) and the counterpart that they were facing. This does not mean that negotiators neglect analysis. They plan for the negotiation, they try to retrieve information about the counterpart, and they analyze the negotiation situation that they are in to identify which heuristics would be ecologically rational to use. Skilled negotiators understand that analysis *and* heuristics can both be valuable tools to reach better negotiation outcomes.

From assembly lines to executive boards, teams do most of the work and make most of the decisions in contemporary organizations. How, then, can organizations build effective teams? With their company positioned as a leader in the technology sector, senior management at Google thought that they had a unique advantage in answering this question: data on their employees, and lots of it. With this goal, Google launched Project Aristotle in 2012. The project team gathered an enormous amount of data—down to who had lunch with whom—on 180 Google teams and conducted in-depth interviews with all of them. The original idea of the company's executives was that the best team would be formed by identifying the most capable people for a job and putting them together. Yet, no matter how the project team looked at the data, they found no evidence that the composition of a team matters. As summarized by the project leader Abeer Dubey, "We had lots of data, but there was nothing showing that a mix of specific personality types or skills or backgrounds made any difference. The 'who' part of the equation didn't seem to matter."[1]

The team had to look for ideas beyond their own data and soon found inspiration from a study published in the journal *Science*.[2] In that study, researchers analyzed a large number of small teams and pinned down two common characteristics of good teams. The first is empathy: members of good teams are generally more skilled in sensing others' feelings and resonating with them. The second is a simple communication rule: each team member speaks for a roughly equal proportion of time, ensuring that everyone's voice is heard. Informed by this finding, the Project Aristotle team reexamined their data and, sure enough, found the existence of these two characteristics in effective Google teams as well. The latter characteristic is essentially the 1/N heuristic applied to team communication:

1/N in team communication: Give each member a roughly equal amount of time to speak.

Although members of a work team can differ in many aspects, treating them as equals gives each member a sense of belonging and respect, motivates them to contribute voluntarily and actively to teamwork, and helps maintain a harmonious relationship within the team. This in turn facilitates collaboration among members, enhancing team performance.

The experience of Project Aristotle illustrates that without knowing what to look for, searching in big data alone is unlikely to deliver. Guided by research findings on teamwork and collaborative communities, in this chapter we explore simple heuristics that teams and communities can rely on to overcome obstacles and accomplish their goals.

Heuristics for Effective Teamwork

Teams can be defined as two or more individuals with different responsibilities who interact with and depend on each other to work on tasks and achieve common goals.[3] The findings of Project Aristotle suggest that how team members interact with each other is perhaps at least as or even more important than who is on the team, assuming that all members meet a satisficing level of qualification. Besides using 1/N, teams can develop a toolbox of heuristics to foster collaboration and boost the quality of interactions. The toolbox includes some of the heuristics introduced in the previous chapters, such as *first listen, then speak, imitate-the-successful, reciprocity,* and *tit-for-tat.* We next describe several additional heuristics that can help teams work together effectively.

For newly formed teams, the organizational psychologist Eduardo Salas and colleagues recommend the *early wins heuristic.*[4] Specifically, they advise firms to assign some simple tasks to the team so its members can quickly experience a sense of accomplishment as a unit and as a result develop a strong feeling of belonging to and belief in the team. Another piece of advice concerns the fit between team leadership and team task: when the task is routine and relatively easy, leadership by a single person is likely to be more ecologically rational, whereas when the task is challenging and requires a lot of coordination, distributing leadership responsibilities among multiple team members likely works better.

Teams often engage in multiple tasks, and when they do, both team leaders and team members can be overwhelmed, losing track of what is going on. Joel Spolsky, the founder of several technology companies (e.g., Stack Overflow), developed a *rule of five* for team task management.[5] Specifically, during meetings, he wanted to hear from each of his teams about five things: two tasks that the team was currently working on, two tasks that they planned to do next, and one task that people might expect them to be working on but they were not actually planning to address. In this way, Spolsky and his teams could prioritize tasks properly, remain focused, and communicate better.

Teamwork requires collaboration from team members, which often takes time to develop; it also benefits from fresh ideas that sometimes are best generated by bringing in new members. Using research collaborations as examples and having analyzed more than 90,000 academic papers, the management scholar Brian Uzzi suggested using this rule to form a productive collaborative team: "You might want 60–70% of a team to be incumbents, and 50–60% repeat relationships. That gets you into the bliss point across four very different scientific fields."[6] The rule provides guidance on how to balance experience and growth, making it valuable for managing many other types of teams, such as sports teams and executive boards.

In addition to the right balance between incumbents and new team members, teams need to be the right size. With too many members, team coordination can become unwieldy; with too few, teams may not have enough resources to complete a task. Rather than trying to optimize using complex formulas, business practitioners and researchers alike recommend simple heuristics for determining team size. For example, Jeff Bezos is credited with having coined the *two-pizza rule*.[7]

Two-pizza rule: Teams should not exist if they cannot feed themselves with two pizzas.

This is similar to the *table rule* by the product designer Erin Casali: the perfect team size is the number of people that can sit around a table without breaking into multiple conversations.[8] Both rules put team size in the single digits, which is consistent with the observation that some of the most powerful teams in the world are small committees, such as the US Supreme Court, the Politburo of the Chinese Communist Party, and the executive boards of Fortune 100 companies. The rules are also supported by research. In a study of 329 work teams in both for-profit and nonprofit organizations, the organizational scholar Susan Wheelan found that teams consisting of three to eight members were more productive and better at negotiation and conflict management than teams with nine members or more.

Why Do Small Teams Work?

Teams are usually formed for two reasons: diversity among team members in knowledge, skills, and information, and coordinated efforts from multiple individuals. Both enable teams to accomplish what is difficult or impossible for a single person to do. As a team gets larger, it will generally become more diverse, which benefits performance. However, larger teams also tend to have more difficulties in coordination, communication, and cohesion maintenance, which affect performance negatively. Thus, an "ideal" team should have neither too few nor too many members but somewhere in between. This can be explained by the theory of "single-peaked preference functions" that the psychologist Clyde Coombs found underlying many human behaviors.[9]

In a single-peaked function, as an independent variable (e.g., team size) gets larger, the value of a dependent variable (e.g., performance) increases, reaches its peak, and finally decreases. Inspired by approach–avoidance conflict in motivation studies, Coombs showed that the increment of the independent variable often has two opposing effects on the dependent variable, one good and monotonically increasing (e.g., diversity) and the other bad and monotonically decreasing (e.g., coordination difficulty). This tug-of-war leads to a single-peaked function, which is guaranteed to occur when good things satiate (i.e., have a diminishing return) and bad things escalate (i.e., get increasingly worse). Figure 8.1 illustrates this dynamic in the context of team size.

Research on the wisdom of select crowds provides further evidence supporting small teams. Different from work teams, a *crowd* may consist of members who do not know or interact with each other and whose opinions are aggregated by some voting or statistical rules instead of discussion.

The psychologist Albert Mannes and colleagues analyzed data from the Federal Reserve Bank's Survey of Professional Forecasters that included nearly 16,000 forecasts of seven economic indicators.[10] They compared the forecasting accuracy of a whole crowd, which had all individuals making a specific forecast and an average size of thirty-five, with that of a select crowd, which consisted of only the top k individuals in the whole crowd according to their

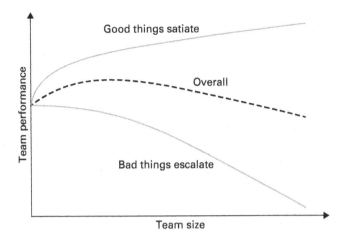

Figure 8.1
Team performance as a single-peaked function of team size. As a team becomes larger, its diversity increases, but this increase has a diminished return (i.e., good things satiate); in the meantime, coordination among team members becomes increasingly more difficult as more members are added (i.e., bad things escalate). The joint effect of these two opposing forces is a single-peaked function between team size and a team's overall performance (middle curve).

past forecasting accuracy. They found that a small select crowd of five to nine members made the most accurate forecasts overall. Further analysis indicated that such crowds performed the best in tasks with an intermediate dispersion of expertise among the members and an intermediate correlation between members' forecasts, two characteristics found in most real-world tasks.

This study focused on judgments, such as predicting next year's unemployment rate. How about decisions? One of us (Luan) collected data in two tasks, predicting the winners of games played in the National Football League (NFL), the professional American football league, and choosing which city of a pair has the larger population.[11] The data contained 2,816 NFL games predicted by an average of sixty-eight sports journalists for each game and 1,063 city pairs chosen by an average of fifty-eight laypeople for each pair. Following the same procedure as in the forecasting study in forming a select crowd, we found that compared to a whole crowd, a small select crowd of around nine members made the same or a higher number of accurate decisions, and the larger the dispersion of individual accuracy, the more likely a small select crowd would outperform a whole crowd.

In both studies, an equality heuristic was used to aggregate judgments (unit weighting) or decisions (tallying). In another forecasting study, researchers developed a complex method to calculate the "optimal" weight for each member of a large crowd on the basis of members' past performance. They found, however, that the highest accuracy in new forecasting tasks was achieved by first forming a small crowd of the top six performers and then weighting their estimates equally.[12] The NFL-game and city-pair study shows a similar result. These results suggest two simple rules for building effective teams or crowds: keep them relatively small (within single digits) and rely on equality heuristics for opinion aggregation and communication.

Dealing with "Bad Apples" on a Team

An advantage of small teams is that they are less likely to have "bad apples," members who are free riding or toxic. However, what should be done when there is a bad apple on the team? One-reason heuristics can help. Unethical behaviors are often cited as the single reason to expel a person from a team or an organization. Consider the case of Scott Thompson, who was named CEO of Yahoo! in January 2012. Before joining Yahoo!, Thompson was the chief technology officer of PayPal and well liked by both management and employees. Yahoo! had high hopes that he would steer the company to a better future during a period when the company was struggling. The hope was extinguished in merely four months, after Thompson was found to have lied about his academic credentials: instead of holding a bachelor's degree in both accounting and computer science, as he had claimed, he actually had only a degree in accounting.[13] Although Thompson's track record and experience exceeded those of most degree holders in computer science, this seemingly innocuous lie cast doubt on his moral character and was sufficient to fire him.

Firing a capable executive because of a lie about their academic record may seem excessive. From a utilitarian perspective, making a decision involves converting the good and the bad of an alternative into a single metric and choosing the option with the highest score. This implies that everything has a value, and one bad quality can be compensated for by one or more good qualities. However, there are qualities, such as integrity, that many people regard as absolute and too important to put a price tag on and trade off. The psychologist Philip Tetlock calls these "sacred values." People are outraged by those who violate or even contemplate violating these values (e.g., trading one's own child for money or knowingly not recalling a dangerous product) and often do not hesitate to sever ties with the violators, if not punish them further. As Tetlock put it: "Opportunity costs be damned, some trade-offs should never be proposed, some statistical truths never used, and some lines of causal/counterfactual inquiry never pursued."[14]

Committing offenses against sacred values can be a single, sufficient reason for ending a relationship. However, in team interactions, lesser offenses, such as not fulfilling a promise or delivering poor work, are more typical. In these cases, how does one decide whether to forgive the offender? In a study by one of us (Luan), we asked participants to recall an instance where they felt hurt or harmed by others in the past six months, to tell us whether they had forgiven or wanted to forgive the person, and to rate the offender on three cues that previous research had shown to be highly related to forgiveness decisions: intention (whether the offense was done on purpose), blame (whether the offender directly caused the harm or could have prevented it), and apology (whether the offender offered a sincere apology).[15]

We found that fast-and-frugal trees with different exit structures could explain well how different participants made their forgiveness decisions. Figure 8.2 shows two such trees: one is quite "liberal" (e.g., requiring little to forgive) and the other is more "conservative." Which type of fast-and-frugal tree a participant adopted corresponded well to their judgments of the relationship value and the potential exploitation cost of the offender. For instance, participants decided with a liberal tree when the relationship value was high (e.g., one's boss) and with a conservative one when the exploitation cost was high (e.g., an abusive romantic partner). Note that on the conservative tree, an intention to harm is considered a violation against a sacred value and leads directly to a "not forgive" decision, but the same is not so on the liberal tree.

In social interactions and team settings, the *lack* of a quality or behavior (e.g., an apology) often functions as a one-clever-cue heuristic for exclusion. In a scene of HBO's hit show *The Sopranos*, the mafia boss Tony Soprano is playing poker with members of his crew. Tony makes a bad joke, and the camera shows in slow motion how all the crew members feign laughter, pretending to enjoy the joke, except for an old mafia captain nicknamed Feech. Tony infers two things immediately: first, he is still the dominant figure in the crew, feared and respected by most of his guys; and second, the fact that Feech does not laugh at his bad joke confirms his suspicion that Feech never truly regarded him as the boss. Some days later, Tony instructs his associates to set Feech up on a burglary charge that sends him to prison.

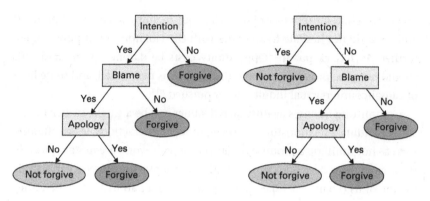

Figure 8.2
Two fast-and-frugal trees for deciding whether to forgive an offender. A person who adopts the left tree forgives more easily and is thus quite liberal, whereas a person who adopts the right one is more conservative, requiring at least the absence of intention to forgive. Based on Tan, Luan, and Katsikopoulos (2017).

Managing Virtual and Crowdsourcing Teams

The three authors of this book reside in three countries on two continents. Video conferencing and file transfer tools provided great assistance to our collaboration, as they do for countless virtual teams. *Virtual teams* are teams with geographically dispersed members who rely on communication technologies for collaboration. With the development of high-speed Internet and advanced collaborative tools, the continued globalization of business, and the challenges brought by natural disasters and pandemics, virtual teams have become quite prevalent. Although many heuristics that facilitate face-to-face teams' performance are applicable to virtual teams, these teams generally experience more difficulties in communication and trust building. These novel challenges require novel solutions.

A commonly recommended rule is that *whenever possible, members of a virtual team should meet face-to-face.*[16] This rule is particularly useful for newly formed teams with unfamiliar members, but it applies to teams generally because face-to-face meetings provide information and cues critical for communication that are often missing in virtual interactions. Other useful rules for virtual teams include (1) *start the work right away*, to avoid delay in production due to the extra time needed for communication and solving technical issues; (2) *acknowledge overtly that you have read others' messages*, to make sure that common knowledge is properly distributed among team members;

(3) *be explicit about what you are thinking and doing,* to lower ambiguity and second-guessing, which are detrimental to a team's cohesion; and (4) *set deadlines and stick to them,* to increase levels of perceived accountability and trust among members. Members of virtual teams instructed to follow these rules developed more trust in their teams and liked other members more than those who did not know about these rules.[17]

Simple rules are also useful for a different form of technology-enabled collaboration: crowdsourcing. In chapter 6, we saw how LEGO used an incentive plan that included royalty sharing to motivate lay designers to remain in a crowdsourcing team and contribute. Other crowdsourcing efforts rely on unpaid volunteers. Consider Wikipedia, probably the most well known crowdsourcing effort on the Internet. From the side of management, Wikipedia has used a *consensus rule* to decide whether an edit on a page is accepted.[18] Specifically, an edit has presumed consensus until it is disputed. Once a dispute occurs, the editors are advised to first seek compromise by repeated editing or discussion among themselves; only when a compromise fails to emerge do outside interventions, such as third-party opinions and arbitrations, kick in. This set of rules embodies the decentralized decision-making approach that makes Wikipedia function well with relatively few resources at its disposal. As for the crowd editors of Wikipedia pages, Darren Logan and colleagues recommended ten heuristics for editing, including *do not infringe copyright; cite, cite, cite; avoid shameless self-promotion;* and *share your expertise and don't argue from authority.*[19] These rules encourage honesty, authenticity, and humility, enhancing the perceived trustworthiness among the editors and of the page content.

Smart Heuristics Protect against the Tragedy of the Commons

Individuals and teams exist and operate within larger communities. For a community to thrive, collaboration between different units is key. This requires the units to combine their efforts, coordinate their actions, and on some occasions, make sacrifices in the pursuit of a collective goal. At the same time, different units also compete against each other for limited resources, such as attention, power, and market share. How to balance this competition–collaboration conflict has been a challenge since time immemorial. Robert Axelrod's research on the repeated prisoner's dilemma game shows that the tit-for-tat heuristic is a simple yet very powerful strategy that can sustain collaborative relationships and guard against destructive behavior (see chapter 7).

Elinor Ostrom, winner of the 2009 Nobel Memorial Prize in Economic Sciences, dedicated much of her research to studying how communities avoid the *tragedy of the commons*: the danger that individual units overuse a common resource until it is depleted. The common resource could be water used for growing crops, fish harvested from a lake for food, or clean air to breathe that is polluted by factories. To prevent the tragedy from happening, some form of mediation from a central governing agent seems inevitable. This approach, however, can get very complex and inflexible and can be inconsiderate of the peculiarities of local situations—in other words, ecologically irrational. Believing in the emergent wisdom of human groups, Ostrom came up with a different approach, trying to understand how real-world communities solve commons problems and to learn from successful cases. In the book *Governing the Commons*, she described effective resource management practices that are found in cultures ancient and new.[20] Much of the discovered wisdom can be summarized as smart heuristics. The following are two examples.

The Wintering Rule

Törbel is a mountain village located in the Valais canton of Switzerland. For centuries, the villagers have planted grains, vegetables, fruits, and hay for winter fodder on their private lands while logging and raising family-owned cows in communally owned areas. Because cheese products are the main source of their income, managing cows grazing on the meadowlands in the summertime has always been an important issue. A document dating back to 1517 stated a simple rule regulating the access to the meadowlands.

> *The wintering rule: No citizen can send more cows to the alp than they can feed during the winter.*

The rule is transparent, fair, and easy to implement, as cowherds always keep a watchful eye on each other. Anyone who is caught sending more cows than what the rule permits will be fined severely. The wintering rule was still in place in Törbel and many other Swiss villages until at least the 1990s.

Turn-Taking

In the 1970s, the Alanya fishery in Turkey was facing a serious crisis. Years of unrestrained fishing had caused hostility and conflict among the fishers, as they all wanted to fish at the best spots. This had in turn reduced the fish stock and increased the operation cost and the uncertainty of harvest for each fishing boat. Members of the local fishing cooperative decided to put a stop to this chaos and, after more than ten years of trial and error, came up with a set of rules that can be summarized as a form of turn-taking.

Turn-taking: Appropriators of a common resource or property take turns accessing the resource by a consensually agreed allocation schedule.

In their case, the rule was implemented during the fishing season as such: in September, all usable fishing spots were named and listed and then randomly assigned to eligible fishers. Every day, each fisher moved east to the next location from September to January and moved west from January to May. This gave all the fishers an equal opportunity to access the fish stocks, which migrated from east to west between September and January and reversed their migration from January to May. This rule is again transparent, fair, and easy to implement. Consequently, the fishers stopped fighting and even restrained themselves from overcatching, as the harvests were now stable and sufficient for everyone.

Other allocation rules built around turn-taking were found in villages in Spain and Nepal for irrigation, and in Japan for harvesting wild plants. Beyond the commons problem, turn-taking has been used widely for conflict resolution and division of labor, such as serving customers at business venues; talking in interviews, debates, and negotiations; child custody for divorced parents; and job assignments within an organization.

Since its establishment in 1946, the United Nations Security Council has rotated its president among the fifteen member states every month; three senior executives at the Chinese technology giant Huawei have taken turns serving as the company's CEO for a term of six months each since 2011; and beyond leadership positions, many firms have job rotation programs that send employees to work in different departments or branches for a set period of time. Among all the possible benefits of turn-taking, distributive fairness is probably the most important. It increases participating members' trust in the organization or the community, and in turn their willingness to cooperate.

Drawing Lots

As in the case of the Alanya fishery, turn-taking is frequently paired with random draws. In his book *The Luck of the Draw*, Peter Stone gives numerous examples of decision making by drawing lots, including the starting position in games and contests, the allocation of medical and other scarce resources, the selection of government officials, the admission of college students, and the hiring of employees.[21] During the 2022 midterm elections in the US, a candidate was elected mayor of the city of Butler, Kentucky, by winning a coin toss over his competitor after each received the same number of votes.[22] Indeed, twenty-eight US states support using a coin toss or some other form of lots drawing to decide election outcomes in the rare cases of equal votes.

Decision by drawing lots is typically applied in situations where any reason-based decision model will likely be judged by some as unjust and met with resistance. In decisions about resource or responsibility distributions, drawing lots is similar to the $1/N$ heuristic, in that it gives equal opportunity to each involved party and is impervious to competing interests. This makes acceptance of the outcomes easier, prevents disputes, and maintains harmony in a community.

Moving Forward

Ostrom suggested that human societies develop various norms, such as turn-taking and lots drawing, to regulate collaboration, and as when learning language, people learn the "grammar of norms" at a young age, reciprocating prosocial behaviors and punishing defections. Similarly, for a team or an organization to thrive, it is important to establish norms early to facilitate the bonding of its members and encourage collaborative behaviors. Norms in the form of simple, transparent rules are particularly helpful because they can be communicated, learned, and revised easily, and more important, they are robust against uncertainty.

From hunting parties to virtual crowds, the forms of human groups evolve constantly. With developments of technologies in virtual and augmented reality and neuroscience, new means of communication and types of teams will undoubtedly emerge in the future. That said, as proclaimed famously by Steve Jobs, "Technology is nothing. What's important is that you have a faith in people, that they're basically good and smart, and if you give them tools, they'll do wonderful things with them."[23] Smart heuristics distilled from years of human experience shall continue to empower teams to do wonderful things, so long as people are still part of a team.

The more than eighty scientific institutes under the umbrella of the Max Planck Society are preeminent sites for basic research in Europe. Their success is based on three heuristic principles. First, research is built around a person, not a field. When searching for a new director, the Max Planck Society does not select a field and look for the best person that it can get in this area, as most universities do. Instead, it looks for an eminent researcher who is then absolutely free to create an agenda or a new field. This is called the *Harnack principle*, named after Adolf von Harnack, the first president of the Max Planck Society, appointed in 1911.[1] The second principle is to plan for the long term and take risks. To enable risk taking, directors are provided with the necessary resources until they retire, freeing them from dependency on short-term grants and the opinions of the average reviewer. This long-term funding reflects an unusual amount of trust. Third, because new ideas do not respect traditional disciplinary borders, a premium is placed on interdisciplinary research.

When one of us (Gigerenzer) was offered a position as a director at the Max Planck Institute for Psychological Research, he faced a leadership challenge. The principles of the Max Planck Society enable innovation, but they do not provide specific guidelines for running a research group. How could a culture of trust be created that nurtures friendly yet rigorous discussion, risk taking, mutual learning, and a healthy error culture? He designed a set of heuristic principles to start the Adaptive Behavior and Cognition (ABC) research group:[2]

- *Common topic/multiple disciplines.* Define a common topic—heuristic decision making under uncertainty—and bring together researchers from psychology, behavioral economics, management, machine learning, mathematics, behavioral biology, and other fields in problem-oriented research.

- *Open culture*. Create a culture of critical but respectful and fact-oriented discussion without regard for hierarchies. Make sure that criticism is leveled not at persons, but at ideas.
- *Spatial proximity*. Locate the entire group on the same floor, keep office doors open, and have daily tea time at 4 p.m., all to encourage personal and professional conversations.
- *Temporal proximity*. Bring all the members of the research group in at the same time to create a level playing field and reduce status differences between the "old guard" and newcomers.

These rules contributed to successfully starting the group. To maintain an open culture, the rules needed to be adapted and new ones introduced. For instance, temporal proximity is important at the start but would be fatal afterward. The culture of a group is passed on by implicit learning, and it lives on even if none of the initial members are still there. It would be lost, however, if all members were replaced simultaneously by a new group. Spatial proximity, in contrast, remains essential for maintaining an open culture. When the ABC research group grew to some forty members consisting of associate and assistant professors, postdocs, doctoral students, and support staff, the architect proposed a new building for the newcomers. To sustain spatial proximity, the director vetoed the proposal and the existing building was extended horizontally so that everyone could still work on the same floor. To uphold a culture of innovation over time, additional heuristics such as the following proved to be effective:

- *Make sure to include a contrarian*. To protect members from falling prey to groupthink, a person who dares to question the group's and the director's wisdom, insists on evidence, and plays the devil's advocate is essential. More than one contrarian could be even better. Such individuals can occasionally be frustrating but actually provide great value.
- *The cake rule*. If a paper is published, the first author brings cake for the entire group. Note that whereas some university departments pay researchers a cash bonus for publishing, the cake rule takes the opposite approach. Instead of the researcher getting a bonus, the researcher rewards everyone else. This reversal of incentives acknowledges that most ideas were inspired

by the entire group. And it emphasizes in-group sharing rather than competition, which is necessary when conducting research as a team.

The toolbox of leadership heuristics needs to be adapted to specific organizational contexts. Whatever heuristics are in a particular leader's toolbox, these set the culture of a group, the tone of communication, and the possibility of innovation and growth. Ecologically rational leadership heuristics are crucial not only for running research groups, but for managing teams, organizations, and even nations. Despite the proven benefits of heuristic decision making—speed, frugality, accuracy, and transparency—most leadership theories surprisingly do not focus on how leaders use heuristics to make effective decisions. To ground our subsequent discussion of leadership heuristics, let's briefly review these theories.

A Very Brief Overview of Leadership Theories

Evolutionary leadership theory traces humans' evolved strategies for leading and following to our prehistoric past. It suggests that many of these strategies appeared during the Pleistocene era, when humans lived in small bands of around 100 seminomadic hunter-gatherers in arrangements that were largely egalitarian.[3] These groups had no formal leadership roles, and leadership was "fluid, distributed, and situational" and accorded to the individuals with the greatest skills and expertise in a particular domain, from hunting to peacemaking.[4] Leading and following mutually evolved to address social coordination problems. For the problem of where to go hunting, for instance, having a leader who was an experienced hunter allowed this decision to be made quickly and accurately.

This changed with the development of larger communities and the institution of bureaucracies and their structural, formal leader roles, which eventually turned into hereditary leadership and aristocracy. Early twentieth-century "great man" leadership theories reflect this perspective and suggest that great leaders have outstanding, genetically based traits such as dominance or intelligence. Think of Jack Welch, Steve Jobs, and Elon Musk. According to this perspective, leaders were male and were born, not made; if you were a woman or did not have the right genes, your destiny was to be a follower.[5]

When great man theories went out of vogue, researchers tried to find personality traits that characterize a true leader. The most well-known personality model is the Big Five, which consist of the five broad personality traits of conscientiousness, agreeableness, emotional stability, extraversion, and openness. Yet the correlations between leaders' Big Five values and followers' job satisfaction, their satisfaction with the leader, their perception of the leader's effectiveness, and the performance of the leader's group turned out to be in the zero-to-small range.[6] Moreover, most of these studies only correlate variables without predicting future performance based on the Big Five, inviting overfitting and inflated estimates.

Just as influential personality psychologists criticized the trait approach for neglecting the roles of the processes and situation, leadership researchers recognized the futility of trying to understand leadership from a trait perspective. Contingency theories of leadership developed models that prescribed how leaders should make decisions depending on situational factors.[7] Vroom and Jago's widely known contingency model, for instance, posits that leaders should first ask themselves whether the decision is important; if not, they should ask whether team commitment to the decision is important; then if not, they should choose an autocratic decision-making style and make the decision without involving the team.[8] These theories have since fallen out of fashion, at least in academic research. A key difficulty was establishing the various contingencies that influence how leaders should make decisions.

The focus turned next to leadership styles, or patterns of behaviors that leaders use consistently in their interactions with others, such as transactional, transformational, laissez-faire, or authentic leadership. Leadership styles generally correlate more strongly with leaders' effectiveness than do the Big Five. However, when considering only studies with more rigorous designs, where data are collected from multiple sources (leaders *and* followers) and longitudinally over time, the correlations are still small.[9] Nevertheless, leadership styles remain highly popular, as the countless books on transformational and authentic leadership attest to. Yet several stubborn problems persist. First, given that leadership styles are conceptualized at a relatively high and abstract level, how effectively do they guide leaders in navigating their daily challenges? For example, a transformational leader is supposed to exert "idealized influence," but what does that mean concretely in the day of a busy leader? Second, context is neglected. Should leaders always be transformational or always be transactional? The principle of ecological rationality, as well as common sense, suggests that different situations call for different behaviors. But theories on leadership styles provide no guidance on when each style should be used.

Other leadership theories similarly offer limited practical advice on the day-to-day challenges that leaders face.[10] Consider the *romance of leadership* view, according to which leaders are actually not that influential; instead, followers' romantic perceptions create larger-than-life images of leaders as almost single-handedly shaping and changing the course of organizations, nations, and history. Or consider theories of emergent leadership that seek to describe how leadership emerges from the interactions of people. All these perspectives capture something important about the complex phenomenon of leadership, and yet the question remains: How do they offer practical advice to people trying to lead? Viewing leadership as decision making and studying concrete leadership heuristics addresses this issue.

Leadership as Decision Making

An anecdote from the American botanist David Fairchild recalls US president Herbert Hoover's reaction to Fairchild talking about a planned expedition to search for novel plants. Hoover asked: "Do you have to make decisions on this trip? If not, I'd like to come along. I'm tired of making decisions, one after another all day long. My view of Heaven is of a place where no one ever has to make a decision."[11]

The leading management thinker Peter Drucker considered decision making the quintessential executive function: "To make decisions is the specific executive task. . . . To make decisions that have significant impact on the entire organization, its performance, and results defines the executive."[12]

Similarly, the seminal work of Simon and March established decision making as a key leadership function.[13] Indeed, most leaders are involved in numerous decisions, some smaller and some larger, each and every day. How, then, do leaders actually make decisions? How do they decide, for example, whether to allocate limited time to a meeting with an employee or a call with a potential customer, whether and how to criticize an employee for substandard work, and whom among a team to promote? To make these and many other decisions, leaders draw on their adaptive toolbox of heuristics.

Leaders' Adaptive Toolbox of Heuristics

Good leadership consists of a toolbox full of rules of thumb and the intuitive ability to quickly see which rule is appropriate in which context.[14] Mitch Maidique, former president of Florida International University, argued that leaders draw on their personal toolbox, which contains a kaleidoscope of experiential rules of thumb. These rules are smart heuristics that leaders learn over time from their own experience or from observing others. According to Maidique, such rules of thumb "are the coin of the realm for experienced CEOs and C-level executives."[15]

Maidique's interviews with twenty CEOs from large US companies revealed numerous leadership heuristics.[16] For example, the billionaire businessman Micky Arison, chairman and former CEO of Carnival Corporation (the world's largest cruise operator) and owner of the professional basketball team Miami Heat, shared this leadership heuristic:

Hire well and let them do their jobs.

This heuristic is similar to the Max Planck Society's Harnack principle: build a research group around a world-leading researcher and give this person absolute freedom to pursue the research agenda. The principle builds trust and encourages quality. Hiring the right people is important, and so is promoting them. Many companies use the following heuristic:

Promote from within.

This heuristic can help ensure competence and adherence to corporate values. People from within the company are observed over a longer period of time, providing a more reliable assessment of their competencies and values. Other leadership heuristics relate to how leaders interact with their teams:

First listen, then speak.

We have already encountered this rule in a different context in chapter 3. When leaders apply this rule, they encourage their staff to share their ideas and opinions openly. In contrast, leaders speaking first encourages conformity with whatever the leader says. This rule is not limited to business leaders. Airplane pilots, for example, are taught to apply this rule during emergency situations to obtain the crew's frank, unintimidated opinions.[17]

Constructive and Destructive Leadership Heuristics

Although leadership heuristics come in many shapes and colors, Maidique differentiates two basic types: constructive and destructive rules of thumb. Constructive heuristics can be further divided into two groups, depending on their level of generalizability: contingent and portable. Contingent heuristics are true specialists: they function in a narrow environment only, such as a specific company in a specific industry under specific market conditions. Portable heuristics, in contrast, are more generalizable, even though they generalize only within specific contexts. An example is this one-clever-cue heuristic used by Ray Stata, the former chairman of Analog Devices:

If a person is not honest and trustworthy, the rest doesn't matter.

Another example of a portable rule is from Bill Amelio, the former CEO of Lenovo:

Build a team of people you can trust.

These rules are generalizable in the sense that one can use them in different scenarios and over a long period of time.

Not all heuristics are used to the benefit of the organization, its customers, or society. Leaders may have questionable goals and use simple rules to achieve them. Maidique calls these *destructive rules of thumb*. For instance, Jeffrey Skilling, the former CEO of Enron, had this rule: "Insider trading is permissible." And the infamous investment fraudster Bernie Madoff's implicit rule of thumb underlying his (and every other) Ponzi scheme was "Defraud your customers until they catch up with you." Heuristics are not good or ethical per se. It all depends on the goals that leaders pursue with them.

Proverbs as Inspiration

Given the important role of political, military, and business leadership in human history, it is no surprise that both philosophers and folk wisdom provide leadership advice. This advice often comes in the form of proverbs. The organizational scholars Li Ma and Anne Tsui argue that these proverbs continue to influence or are reflected in current leadership.[18] Although not heuristics themselves, these proverbs can be considered the ground that inspires heuristic leadership strategies. For example, a famous statement on leadership from the traditional Chinese philosophy of Daoism is that "governing a large state is like cooking a [pot of] small fish."[19] One interpretation of this proverb is that in certain situations, it is better to take no action, suggesting the following heuristic:

Don't stir the pot; do nothing.

The rationale behind this heuristic is that by not "stirring the pot" (not taking action), a leader can avoid doing harm, let people take care of themselves, and lead via quietude.

Traditional proverbs can be valuable sources of inspiration for heuristics. At the same time, there are so many proverbs, making it difficult to know how to pick the appropriate one. This is the question of the ecological rationality of leadership heuristics.

Ecological Rationality of Leadership Heuristics

Not only are there many proverbs, but some of them seem to contradict each other. For instance, proverbs suggest both that "opposites attract" and "birds of a feather flock together"; that "the early bird gets the worm" and "all good things come to those who wait"; and that "clothes make a man" and "you cannot judge a book by its cover." Which advice should a person follow? The key, as always, is to consider the context. For example, the relatively passive style of leadership recommended by Daoism could be intended for hereditary leaders who prefer stability and wish to avoid change. Such an approach to leadership, however, would not work during times of crises or in turbulent conditions.[20]

Generally, leaders would do well to employ their adaptive toolbox intelligently and flexibly—or with *agility*, to use a currently popular buzzword. Heuristics can become obsolete when the environment changes, and an overly static or mechanistic use of leadership heuristics is doomed to failure. A good example comes from Intel, the multinational microprocessor company.[21] In the early 1980s, Intel was not yet famous for its microprocessors but was known as "The Memory Company" for being the world's leading memory chip producer. The "Intel Way" was to offer a full product line and to use memory products as a test bed for new technologies. However, with the rise of Japanese competitors, Intel faced difficulties and its memory division was making losses, in contrast to its growing market of microprocessors. At this point, Intel's founder, chairman, and CEO Gordon Moore and president Andy Grove met; after a long discussion, Grove famously asked Moore: "If we got kicked out and the board brought in a new CEO, what do you think he would do?" Moore answered: "He would get us out of memories." Grove looked at him and after a long pause said, "Why shouldn't you and I walk out of the door, come back in, and do it ourselves?" As a result, Intel's two leaders adapted their strategies to fit the changed competitive environment: "We don't need a complete product line," and "Microprocessors, not memories, will be Intel's core product line." Soon thereafter, Intel thrived again. Heuristics work well only under appropriate environments; when the environment changes, the heuristics also need to change.

The principle of ecological rationality applies as well to the heuristics covered earlier. For example, the "first listen, then speak" heuristic targets leaders, not followers. If followers first hear the leader's opinion, they will often be tempted to support the leader, even when they disagree internally. The "promote from within" heuristic is more applicable in stable situations. In contrast, "hire from outside" makes sense when fresh ideas and some disruption are needed. The former is used more often when an organization desires incremental development, and the latter when an organization is in crisis and seeks more drastic change. Both approaches come with pros and cons: "Insiders understand their organizations' specific issues, actors, and resources; but they tend to be wedded to the status quo. Outsiders bring fresh perspectives and openness to change; but they tend to be naïve about their new organizations and are prone to missteps as a result."[22] How leaders can learn the ecological rationality of heuristics is a question that we return to in more detail in chapter 13. For now, let's examine a particularly complex leadership task.

Managing Complex Megaprojects

Megaprojects include constructing new airports and holding the Olympic Games. Such projects routinely overrun their original time and budget plans, often by massive amounts; some fail entirely. Consider the case of the Tokyo Summer Olympics in 2020. The official cost estimate at the time Japan made its bid to the International Olympic Committee in 2013 was 734 billion Japanese yen. In 2019, it was already on track to go almost 100 percent over budget, with a projection of 1.35 trillion yen. Then, the "black swan" event of the global COVID-19 pandemic hit, leading to postponement of the

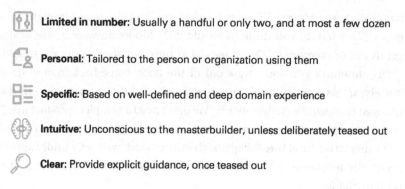

Limited in number: Usually a handful or only two, and at most a few dozen

Personal: Tailored to the person or organization using them

Specific: Based on well-defined and deep domain experience

Intuitive: Unconscious to the masterbuilder, unless deliberately teased out

Clear: Provide explicit guidance, once teased out

Figure 9.1
Features of effective masterbuilder heuristics for managing megaprojects. Effective project leaders intuitively use a few specific heuristics that provide clear guidance and are tailored to the context. Based on Flyvbjerg (2021).

Games to 2021. This added a further estimated 103 billion yen to the price tag, along with many nonfinancial costs. All in all, the final cost at the end of 2021 stood at 1.45 trillion—about double the original budget.[23]

A main reason for such failures is poor decision making. And a key cause of poor decisions is using unsuitable decision heuristics. When, in contrast, megaprojects are completed on time and within budget, project leaders have relied on effective heuristics. The economic geographer Bent Flyvbjerg calls leaders who use such heuristics for project management *masterbuilders*, after the architects of the grand cathedrals built in the Middle Ages.[24] According to his research, masterbuilder heuristics share the features displayed in figure 9.1.

Masterbuilder heuristics vary widely in their details, as they are tailored to different projects. However, some of the heuristics are more broadly applicable. An example is the following rule:

Ensure skin in the game.

Given the numerous interdependencies of multiple parties in megaprojects, it is especially important that each party can be relied on to do their part. If not, dramatic domino effects occur in other parts of the project. If parties do not have "skin in the game" to complete the overall project on time and on budget, chances for success are greatly reduced.

Another more broadly applicable heuristic is to approach megaprojects as one would LEGO constructions:

Use simple building blocks and combine them as needed.

In other words, one should use modular, standardized components rather than trying to create unique, bespoke solutions. Whereas costs and delays for the latter easily spin out of control for complex projects, standardized blocks can be improved over time and costs can be reduced and predicted more easily. Such an approach has a positive learning curve, whereas bespoke approaches tend to have flat or even negative learning curves and economies of scale.

The *Ur-Heuristic*

One of the key simple rules that many masterbuilders use extensively, what Flyvbjerg calls the *ur-heuristic*, is the following:

Keep it simple.

Why does this heuristic work in project management? We might intuitively presume that complex projects require complex solutions. In fact, the more complex the project, the simpler the masterbuilder's approach needs to be. This is particularly true for megaprojects, such as Olympic Games or airport construction. Complexity should be met with simplicity rather than complexity. Here, once again, less is more.

Flyvbjerg draws on the idea of phronesis in his work on masterbuilder heuristics. *Phronesis* is an old Greek term that can be translated as practical wisdom (its Latin translation is *prudentia*, which became *prudence* in English). Ancient Greek philosophers such as Socrates and Aristotle highly valued phronesis, distinguishing this practical form of wisdom from intellect and knowledge. Practical wisdom is based on experience and is action oriented. It is considered a necessary characteristic of people with good judgment. Unlike other forms of wisdom, such as mathematics and logic, which seek to establish general laws, phronesis recognizes the importance of context. It is about how to act wisely in particular situations. Phronesis thus shares many aspects with the concept of ecological rationality. Indeed, the tailoring of the masterbuilder heuristics to the specific person, organization, and task suggests that a key feature is their ecological rationality: Through experience, they were honed for their tasks.

Choosing Leaders

So far, we have focused on the heuristics that leaders use. Humans (and other species) have also developed heuristics for selecting leaders. For instance, using the equality heuristic, each member of a selecting group, such as a committee, board of governors, or the general populace of a country, has one vote; the majority rule then says that the person with the most votes becomes the next leader (CEO or president). The goal of such selection rules may be not only to find the best leader, but also to involve large parts of the community in the election process so everyone feels that they have participated—an important aspect of democracy. Consider the selection of doges of Venice, one of the greatest commercial world powers in the thirteenth to sixteenth centuries. Over multiple rounds of nominating electors and candidates, reducing them by drawing random lots, and repeating this process, a new doge was finally selected. This ensured that many members of the leading families were involved at some point during the process.[25]

In addition to employing such formal selection rules, groups use informal heuristics for choosing leaders. Some of these stem from our evolutionary past and continue to influence who becomes the leader of a group. For example, among prehistoric hunter groups, leading could mean something as concrete as making the first physical move. Research on the first-mover heuristic suggests that individuals in a group indeed tend to follow whoever makes the first physical move, and the simple act of moving together in one direction (whichever direction it is) provides advantages over uncoordinated movement.[26] To the present day, groups still tend to follow the person who speaks up first (and loudest) or takes the first action, and this person becomes the de facto leader of the group.

We also appear to prefer leaders who fit certain leadership prototypes already in use among hunter-gatherers: "big men" who are strong warriors.[27] To exaggerate somewhat, it is as if the stereotypical Neanderthal is leading a team (figure 9.2). The concept of evolutionary mismatch implies that to the extent that environmental conditions have changed, evolved behaviors may no longer be effective or appropriate.[28] For example, during our evolutionary past, eating as much high-calorie food as possible was helpful to prevent starvation and death. However, under conditions where high-calorie foods such as soft drinks and fast food are abundant and cheap, the same evolved tendency leads to obesity and premature death. Similarly, selecting leaders on the basis of physical strength, dominance, or aggressiveness may have been functional under ancestral environments but makes little sense in today's corporate world.[29] The heuristic is no longer ecologically rational.

Figure 9.2
Selecting leaders (typically males) heuristically based on physical strength or aggressiveness is an example of evolutionary mismatch: a lack of fit between evolved behaviors and present-day conditions. Source: Courtesy of Kikuko Reb.

Following a leader offers the promise of benefits that could not be gained otherwise, but also the risk of being taken advantage of by the leader.[30] This is called the *follower's dilemma*. Fairness heuristic theory offers a solution: under conditions of uncertainty, leaders' procedural fairness serves as a helpful heuristic to predict valued future outcomes.[31] When leaders use fair procedures, such as treating everyone equally, employees' uncertainty about future outcomes (e.g., receiving a deserved bonus or promotion) is reduced because the link between performance and the desired outcome is then stronger.

The Five Benefits of Heuristics for Leadership

In chapter 2, we provided four major reasons why people use heuristics: they are fast, frugal, accurate, and transparent. These advantages are highly relevant for leaders. In addition, leadership heuristics can foster healthy organizational cultures, a fifth benefit.

Heuristics Enable Fast and Accurate Decisions

Leaders eschew complex utility maximization in favor of simpler decision strategies, and for good reason. If executives tried to maximize expected utility, much of their time would be spent collecting information, analyzing data with spreadsheets, and making calculations. They would never get all their work done. The management scholar Henry Mintzberg followed executives around as they went through their workday.[32] He found that managers' days are characterized by a high pace, fragmentation, and action orientation. Leaders rarely have the luxury to sit in their offices and quietly analyze problems. Instead, they face constant interruptions, crises, and urgent demands on their time and attention. Mintzberg found that leaders typically just spent a few minutes on a task before moving on to another. To cope with these demands, they need simple rules to make decisions and take actions quickly *and* accurately, based on limited information. Leaders do so by combining their experience with smart heuristics.

Heuristics Reduce Attention Overload

As recently as two decades ago, the area of management information systems was very popular. The aim was to create systems that collect and provide information to leaders for analysis and decision making. At the time, information was considered scarce. Today, we live in a world with *too much* information at our fingertips, thanks to the Internet and big data. As Herbert Simon observed long ago:[33]

> In an information-rich world, the wealth of information means a dearth of something else: a scarcity of whatever it is that information consumes. What information consumes is rather obvious: it consumes the attention of its recipients. Hence a wealth of information creates a poverty of attention and a need to allocate that attention efficiently among the overabundance of information sources that might consume it.

We live in an attention economy.[34] Defining features of this age are the competition for our attention and the widespread feeling of attention overload. Leaders, perhaps even more than many others, suffer from information overload. Much of the available data is noise. It is a challenge to separate the signal from this noise and not to be drowned by the overwhelming amounts of data.

How can leaders deal with information and attention overload? One approach is to outsource decision making to machines and artificial intelligence (AI). At some point, perhaps, we will have organizations that no longer require leaders to make decisions, as AI has taken on this role. Whether this is a realistic or desirable vision remains debatable. In the meantime, frugal heuristics can help leaders reduce information overload by enabling them to make decisions based on limited but valid information and on expertise and intuition developed over time.

Heuristics Foster Transparent Leadership

Transparency is an important expectation for leaders, especially in democratic societies and organizations, but also more generally. Decisions made secretively behind closed doors naturally raise suspicion. Employees and society at large want to—and arguably have a right to—know how decisions are made that affect them. However, many leaders appear reluctant to be transparent about their decisions. Sometimes they have something to hide, such as when they engage in favoritism and political games for their own benefit. But even when they are well intentioned, they may fear being blamed if something related to their decisions later turns out to go wrong.

Thanks to their simplicity, heuristics are prime candidates to foster leadership transparency. This simplicity makes it easier to be transparent about what factors were considered in the decision and how they were processed, thus making a clear connection between input and output. A key advantage of such transparency is that leadership heuristics are easy to teach, learn, and transfer for the benefit of the organization. For instance, work on leader heuristic transfer—defined as "the conveyance of a leader's heuristics or experience-based 'rules of thumb'"—found that the extent of the transfer was positively related to employee creativity.[35] Leadership heuristics could also form the core of leadership development training. Leadership development is an industry with an estimated annual value of several billion dollars in the US alone—yet with limited evidence to show for its value. We believe that this is at least partly because such trainings focus on distal, abstract leader traits and leadership styles rather than on the more proximal, concrete heuristics that smart leaders use. We come back to the topic of teaching and learning heuristics in chapter 13.

Heuristics Shape Organizational Cultures

The heuristics that leaders use shape the culture of an organization, for better or worse. For instance, consider "hire well and let them do their jobs." The first part fosters quality, and the second part a culture of trust. In contrast, leaders who micromanage do not convey confidence in their employees. Similarly, leaders who use "first listen, then speak" signal that they take their employees' views seriously. In contrast, leaders who first tell everyone that what they think is right create a hierarchical culture and induce a fear of speaking up. Leadership heuristics can create an open or a defensive decision-making culture (see also chapter 11).

Heuristics can also define moral behavior. For instance, Nelson Mandela spent nearly three decades in prison for seeking freedom for South Africans of color like him, who were oppressed by the apartheid regime. Yet after being released from prison, he sought not revenge but reconciliation. To do so, he relied on leadership heuristics such as "look forward, not backward" and "forgive, don't seek revenge." Mahatma Gandhi's leadership was guided by this simple principle: "Never react with violence, regardless of the provocation." Countless organizations and communities have been elevated by smart leadership heuristics, just as others have been ruined by ineffective ones. Leaders would do well to recognize the great value of their adaptive toolbox and learn how to select the appropriate leadership heuristic for the task at hand.

Part III

Plate 10

The power of intuition—the ability to know more than we can explain—is a striking phenomenon. Good management decisions are often made "straight from the gut," as Jack Welch of General Electric (GE) once put it plainly.[1] An experienced manager can have a gut feeling that something is wrong with a deal without being able to immediately say what, but the hunch leads to a deliberate search for causes. Intuition is a general mark of expertise not only in management. In a study, seventeen Nobel laureates from physics, chemistry, medicine, and economics were asked how they made their "big leap." The majority explained that their discoveries resulted from switching back and forth between intuition and analysis.[2] Similarly, success in business and management requires *both* intuition and deliberate analysis.

We are equally struck by the increasing unconditional mistrust of intuition in parts of the social sciences. Here, intuition is presented as the enemy rather than the ally of reason. For instance, various dual-system theories pit an intuitive "system 1," which is fast, heuristic, unconscious, and often wrong, against an analytic "system 2," which is slow, logical, conscious, and virtually always right. Relying on intuition can go wrong, of course, but relying on algorithms and deliberate thought is equally prone to error. Nonetheless, many bestsellers, including *Predictably Irrational*[3] and *Nudge*,[4] associate error exclusively with intuition and never with flawed logical reasoning or faulty applications of rational choice theory in situations of uncertainty. Falsely depicting intuition and reason as opposites implies that human intuition is not to be trusted and should be replaced by logic or algorithms.

The mistrust of intuition is nothing new. Albert Einstein noticed it when he said: "The intuitive mind is a sacred gift and the rational mind is a faithful servant. We have created a society that honors the servant and has forgotten the gift."[5] Einstein's insightful comment has been internalized at least by the hard sciences, where—unlike in the social sciences—calling something "intuitive" can indicate great respect.

What Is Intuition?

Here, we are not talking about a sixth sense, God's voice, or the arbitrary decision making of an inept leader. An intuition, or gut feeling, is a judgment with all the following characteristics:

- based on years of experience
- appears quickly in one's consciousness
- whose underlying rationale is unconscious

In other words, intuition is not caprice, but rather a form of unconscious intelligence.[6] For instance, an experienced doctor may sense in a blink that something is wrong with a patient, without being able to fully explain why. There is an interesting similarity with the black-box artificial intelligence (AI) algorithms that we discuss in chapter 12: the processes that generate an intuition are not transparent to the conscious brain. However, the doctor's next step is to follow up the initial intuition and begin systematic medical testing. That is, unconscious and deliberate actions complement themselves, and when the diagnosis is successful, the doctor may begin to understand just what generated the feeling that something is wrong. Importantly, intuition should not be seen as the opposite of deliberate thinking; in important decisions, one has to switch back and forth between deliberate and intuitive judgment. Virtually every important business decision is based on both. Rarely do the data speak for themselves.

Do Executives Make Gut Decisions?

An executive may be buried under a mountain of information, some contradictory, some of questionable reliability, and some shaped by particular interest groups. There is no algorithm to calculate the best decision in such a situation of uncertainty. Yet an experienced executive may have a gut feeling about what the best action is. By definition, the reasons behind this feeling are unconscious. To gain a sense of how many important professional decisions in large companies are based on gut feelings, one of us (Gigerenzer) interviewed thirty-two managers, senior managers, and the executive board of a large international technology services provider. With the help of a top executive who had their trust, they were asked in personal interviews how often important professional decisions that they made were ultimately gut decisions, a definition of which they were given beforehand to ensure comprehension.[7] The emphasis was on "ultimately" because we assumed executives first inform themselves with the data related to their decisions; only if the data is not unequivocal will they rely on their gut feelings. The executives in question were from all levels of the hierarchy: managers, heads of departments, group executives in charge of a branch of the company, and members of the executive board. All responded to the invitation for an interview without having to be asked twice, indicating how important they found the issue.

Not a single executive stated that they never made gut decisions (figure 10.1). Nor did a single one say that they always made gut decisions. Instead, the majority (twenty-four of thirty-two) stated that 50 percent or more of their professional decisions were, after consulting the data, ultimately based on gut feelings. This held true through the entire managerial hierarchy of the corporation. It also holds in many other corporations. For instance, among the fifty top executives of an international car manufacturer, consisting mostly of engineers, everyone said that 50 percent or more of their important decisions were based on gut feelings.[8] The higher the management level and the proportion of engineers in the company, the higher the number of reported gut decisions.

The Fear of Admitting Gut Decisions

The very same executives, however, would never admit to making gut decisions in public. A gut decision leaves the burden of responsibility on the leader's shoulders. In large corporations, fewer and fewer managers are willing to take on this responsibility in fear of negative consequences from stakeholders. Interviews with the executives of the international technology services provider showed that they feel under pressure to provide rational reasons for a decision, and intuition by definition cannot provide any. As one group executive explained, "It's the simple truth that one has to apologize if a decision is not based on 200 percent facts." Another one said: "We are a high-tech company and our leadership expects numbers and facts." In their view, it would likely not go over well if a manager publicly admitted: "I looked at all

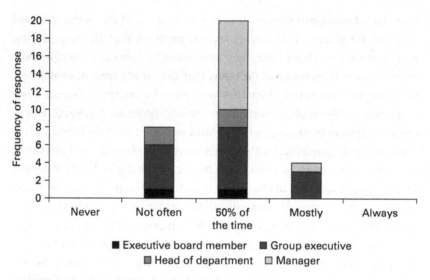

Figure 10.1
How often do executives make gut decisions? Self-reports of thirty-two executives from a large international technology services provider. No executive said that they never or always use intuition, with the majority (twenty) saying that they use intuition about half the time. Based on Gigerenzer (2014).

the facts, but they did not provide a clear answer. Based on my experience, my intuition is to make this decision." To deal with this conflict between intuition and justification, managers resort to two methods of concealing their intuitions: producing reasons after the fact and defensive decision making. These methods show how costly the fear of admitting to making gut decisions is to corporations.

Produce Reasons after the Fact
Instead of presenting the gut decision openly, the executive asks their staff to find reasons after the fact for the intuition, which can take a week or so. With the list in hand, the executive then presents the gut decision to others as a deliberative decision derived from data alone. Rationalization after the fact is a waste of intelligence, time, and resources. It also slows the decision process. In fact, the technology provider in question suffered from problems with slow decision making, both inside the firm and outside when dealing with customers.

Another version of the same strategy is to hire a consulting firm. The consultants will deliver a document full of reasons for the gut decision, not even mentioning intuition at all. This procedure costs even more money, time, and attention. Its ultimate motivation is a leader's reluctance to accept personal responsibility—which is what a gut decision is all about. How often does this happen? Over lunch, we asked the principal of one of the largest consulting firms worldwide whether he would be willing to divulge how many of their customer interactions entail justifying decisions after the fact. His response was, "If you don't disclose my name, I will tell you. It's more than 50 percent."

Defensive Decision Making

The second way to deal with the anxiety of making gut decisions openly is defensive decision making. Instead of acting on the intuitive best option for fear of something going wrong, the manager promotes an inferior option.

> *Defensive decision making: Managers rank option A as the best but nevertheless pursue option B, seen as inferior, to protect themselves in case something goes wrong.*

Consider an experienced manager who has a gut feeling that the company should enter a foreign market with a new product. Yet they do not pursue this option because if it fails, they would be responsible and unable to explain why they promoted the idea. The goal is to protect oneself rather than take risks in the interest of the company. Whereas finding reasons after the fact slows down decision making and incurs unnecessary costs, defensive decision making can directly inhibit innovation and cause companies to forgo lucrative opportunities. It is probably no accident that PayPal was not invented by a large bank and Google was not invented by a large media corporation (see chapter 6).

How frequent is defensive decision making in large corporations? The thirty-two executives at the international technology services provider were asked, "Consider the last 10 important professional decisions in which you participated. How many had a defensive component?" Only seven said "None." One of them, a male executive in his fifties, explained: "I do well if the company is doing well. It's my passionate conviction. Even if this company gave me the pink slip, I would still do the same for the next company."[9]

Yet this ideal type of manager was in the minority. Twelve executives said that they had made defensive decisions in one to three of the last ten cases. One admitted that he had sometimes simply lacked the courage to choose the riskier but more promising option. Others confessed that they were motivated by fear of blame and of compromising themselves by being responsible for an error, which might lead to loss of peer esteem. Ten executives revealed that about half of their decisions were defensive. Several of these justified opting for second-best alternatives because they had no incentives for taking risks—only criticism or punishment if something went wrong. A member of the executive board admitted that every other decision that he made was not in the company's best interest; given the reigning no-risk mentality, he focused on his own career and interests. Finally, two managers, both at the lower end of the company's hierarchy, said that they had engaged in defensive decision making most of the time, in seven to nine out of ten cases. One noted that the company had no error culture and zero tolerance for errors, so his motto was simply "Cover your ass."

A similar number of defensive decisions can be found in other corporations as well as in public administrations.[10] They are a sign of ineffective leadership and a negative error culture (which we discuss in more detail in chapter 11). Defensive decisions occur less frequently in family businesses and owner-led companies. In the culture of family businesses, owners are less anxious about having to justify gut feelings, and their trusted executives worry less about being fired on the spot if something goes amiss. Errors are discussed more often in order to learn from them, plans are made far ahead, and gut decisions are measured by their performance, not by the decision maker's ability to justify them. And the "skin in the game" of family businesses is a strong motive to avoid the costs of appointing consulting firms to camouflage intuitive decisions, or the even larger costs of defensive decision making.

Intuition and Heuristics: The Fluency Heuristic

By definition, the person who has an intuition cannot explain it. To explain how intuition works is the task of research. Consider the surprising fact that the first idea that comes to mind is often the best.

Courses in decision research teach that experienced people carefully compare options, whereas novices jump on the first option that comes to mind. According to the decision researcher Gary Klein, it is the other way around.[11] Klein and his team have slept in fire stations, ridden in M-1 tanks and Black Hawk helicopters, and observed high-stakes decisions in intensive care units. Compare these adventurous sites of *naturalistic decision making* with the safe environment of the psychology lab; in the latter, participants make choices between largely hypothetical lotteries and gambles that they have never encountered before and where everything is known, including the probabilities.

Klein reported that the experts he studies, such as firefighters and emergency room physicians, rarely compare options.[12] Rather, from their experience, a single option comes to mind. This process is known as the *fluency heuristic*, which we introduced in chapter 2. The fluency heuristic has also been called the *take-the-first* heuristic or *recognition-primed decision making*.[13] An expert may pick the first option immediately or mentally simulate the option—that is, imagine it being carried out. If the simulation fails to lead to the desired goal, then the same process is repeated with the second option that comes to mind, and so forth. Relying on fluency is ecologically rational for experts, where fluency correlates with the quality of the alternatives, but not for novices. To make fluency work, the human brain evolved to detect subtle differences in fluency, a requisite ability for applying the heuristic. People can tell the difference between fluency latencies if they exceed 100 milliseconds, and the fluency heuristic predicts individual decisions more accurately the larger the differences, up to 82 percent correct.[14]

As shown in figure 2.1 in chapter 2, the first option that comes to mind to experienced handball players is most likely the best one, and taking more time and generating more options will likely decrease performance. This finding contradicts the hypothesis that there is a general trade-off between making decisions quickly and making them accurately, the speed–accuracy trade-off. This trade-off holds for novices: that is, people who are given tasks they never performed before, as is the case in almost all psychological studies. Yet the trade-off does not hold for experts who can rely on good intuition and whose spontaneous judgment is largely the best choice. This tendency has been reported for firefighters, pilots, and other experts.[15] As the former German soccer player Gerd Müller, one of the top scorers of all time, once remarked about his intuitive play, "If you start to think, you are already a goner."[16]

The fluency heuristic brings home an important insight. Decision making is not necessarily the same as a choice from a set of given alternatives. When making decisions, several options may be considered but are not compared; that is, options are evaluated one by one until one is found to be good enough. And if the situation changes because, for example, a fire has suddenly spread, this option-generating intuitive process is started again. The focus of decision theory on choosing between lotteries has largely obscured this important insight.

The process of intuition here is a combination of the satisficing heuristic—to accept the first alternative that meets an aspiration level—and the fluency heuristic, which orders alternatives according to their validity. Satisficing by itself is mute about the order in which alternatives are met.

How to Block Intuition

The preceding analysis of the fluency heuristic reveals how to block good intuitions of one's professional opponents in competitive situations: Make them think, and make them take more time. The power of the fluency heuristic vanishes if one takes too much time for a decision and considers too many options, which increases the likelihood of choosing an inferior one (see chapter 2). Some people fool themselves by not following the first option that comes to mind and instead searching further. In contrast, those who understand the power of intuition fool their opponents by deliberately using this counterheuristic.

Make your expert opponents think rather than follow their gut.

In 2006, the Olympic Stadium in Berlin was packed with 75,000 fans attending the soccer World Cup quarterfinal game between Argentina and Germany. After extra time, the game was still tied and the penalty shoot-out began—a nerve-wracking situation where the two goalkeepers alternately face five players from the opposing team. All shoot-outs are tense, but this one was special. Before each penalty kick, the German goalkeeper Jens Lehmann studied a slip of paper that he held in his hands. In the end, Lehmann blocked two penalty kicks, eliminating Argentina, and Germany moved on to the semifinals. The media attributed the victory to the information on Lehmann's "cheat sheet."

But that is likely not the case. Consider the final Argentinian shooter, Estéban Cambiasso. A video still available on YouTube shows how Lehmann took his time and studied the sheet of paper. Unknown to Cambiasso, it actually contained no information about him. Cambiasso made his kick, and Lehmann blocked it. It was not the information on the piece of paper that helped Lehmann, but more likely the fact that he made the Argentinian player think about what he should do. Simply by deliberating, Cambiasso swallowed the bait.

The inhibitory influence of thinking on good intuitions has been shown in a number of experiments. For instance, when novices and expert golfers were asked to pay attention to their swing while making a putt, novices did better, but with experts, it was the opposite: their performance decreased.[17] In another condition, novices and expert golfers were given either only three seconds for each putt or all the time they wanted. Under time pressure, novices performed worse and had fewer target hits, which is no surprise. Expert players, in contrast, hit the target more often when under time pressure. The more time an expert has, the more inferior options come to mind and the greater the likelihood that the expert will follow one of these. The general conclusion is that for those highly proficient at a task, thinking too long should be avoided, whereas beginners at a task should take their time in deciding what to do.

The Sacred Gift

As mentioned previously, Einstein spoke of intuition as a sacred gift. Yet underlying the gift are years of hard work and experience with a particular subject. Einstein's remark that we have created a society in which the gift is forgotten remains valid today. In this chapter, we addressed the illusory view that intuition is the enemy of rational thinking, a view embodied in quite a few dual-system theories that frame intuitive, heuristic, and fast thinking as often wrong and pit such approaches against rational, logical, and slow thinking that is framed as virtually always right. That dichotomy forces executives to engage in costly practices to hide their gut decisions, or even make second-best decisions. Fear of admitting to gut decisions is promoted by a negative error culture, in which errors are covered up or punished, and where taking a risk for the company is a dangerous thing. That in turn leads to less innovation, and to a culture in which every new idea needs to be justified; otherwise it is likely to be dismissed. Managers need to overcome the impulse to see intuition and reason as opposites and realize that the two need to work together. How organizations can encourage such a culture is the topic of the next chapter.

11 Creating Smart Decision-Making Cultures

In chapter 10, we saw that managers often resort to two costly actions: producing reasons for gut decisions after the fact, and choosing a second-best option (i.e., defensive decision making). Underlying these behaviors is often a dysfunctional organizational decision-making culture. Because of their importance, we delve more deeply into functional (smart) and dysfunctional decision-making cultures in this chapter.

Decision-making cultures vary widely across time and space. In medieval Europe, for example, it was common to *decide by ordeal* whether someone was guilty, a liar, or a witch. One form was ordeal by combat: two parties in a dispute fought each other, and the loser was considered guilty or liable. A particularly nasty form was *ordeal by water*: A woman accused of being a witch was submerged in water. If she floated, she was judged to be a witch and executed; if she sank, she was judged to be innocent—but died from drowning. In other cultures, important decisions, such as whether to go to war, were made by oracle. It is hard for us now to comprehend how such decision-making cultures could have existed. A culture, as the customs, practices, values, beliefs, and symbols of a nation, society, organization, or group, can look irrational from the outside, but members of the culture generally take it for granted.[1]

Modern cultures are no exception. Consider how decisions are made in contemporary organizations using calculations, spreadsheets, analyses, and reports, sometimes based entirely on algorithms with no human input. To an indigenous person of an Amazonian tribe, this might not make any sense at all. Because culture is deeply ingrained, its influence on organizations can be more powerful than that of carefully laid out strategic plans, as captured in a quote attributed to the management thinker Peter Drucker: "Culture eats strategy for breakfast."

In this chapter, we specifically look at the *culture of decision making* in organizations. That is, we examine the ingrained norms, values, and beliefs— such as the belief that collecting more information is always better—that influence how decisions are made, delayed, or avoided. We ask: How do decision-making cultures differ, and how can organizations develop smart cultures?

Rhetoric Versus Reality in Managerial Decision Making

The culture of decision making is about both how decisions are made and how they are talked about. Managers do not always "walk the talk": They talk of "optimizing" and "maximizing" even when making decisions about large-world problems where, by definition, determining the best course of action ahead is impossible and optimization is an illusion. The rhetoric is about exhaustively searching for information, carefully analyzing the data, considering all possible options, and choosing the best option. In reality, managers rely on a combination of heuristics and analysis. For example, as we described in more detail in chapter 10, most executives in a large international technology services provider indicated that they ultimately made their decisions using intuition.[2] Yet the use of heuristics, consciously or unconsciously (i.e., intuition), is typically not talked about in public.

Consider budgeting. Organizations and governments need to allocate scarce resources over various alternatives. A company could invest heavily in developing a new product or spend the same amount on increasing the market share of an existing product. Complex methods such as net present value calculation and zero-based budgeting attempt to allocate budgets to maximize the return on resources. Yet, by definition, a budget cannot be allocated optimally, as future investment returns are not known at the time of making the decision. As an alternative, decision makers can use heuristics to resolve the allocation problem. One candidate is the recency heuristic: use last year's budget plus or minus delta.[3] Another is $1/N$: divide the budget equally over N employees in the same unit, such as 3M and Google giving their engineers and scientists the same amount of time off to work on independent research projects (as discussed in chapter 6).

Or consider tax rates. For instance, in 2022, Singapore announced adjustments to its tax rates. The goods and services tax increased by 1 percent from 7 percent to 8 percent in 2023 and will go up by another 1 percent, to

9 percent, in January 2024.[4] As it is impossible to calculate an optimal tax increase, a heuristic such as this one can be applied:

Increase taxes cautiously and in equal steps.

This rule reduces the risk of making changes that are so large as to cause a shock to a system that has worked well in the past, and it also allows the government to observe the effects and make adjustments as necessary. Note that the revenue service might have spent considerable time deciding on this rule and whether to increase taxes by 1 percent or 1.5 percent. Again, we see that deliberation and heuristic decision making typically go together.

How organizations talk about and make decisions is part of their culture. In what follows, we describe four dysfunctional decision-making cultures, followed by three functional cultures. These are not mutually exclusive; they can coexist within the same organization and systematically differ across an organization's units.

Rationalization Culture

A disconnect between how organizations talk about versus make decisions can indicate a rationalization culture. We define a *rationalization culture* as one in which decisions are made using heuristics, a fact that is not spoken about, and the decisions are subsequently presented to others—superiors, colleagues, or the public—as if they were arrived at solely by analysis, logic, and optimization. In this way, the heuristic process is "rationalized" (and covered up).

Rationalization can be used to make a decision proposal more convincing, such as when the division of a company wants to develop a new product and needs to obtain the headquarters' support. It is also used to protect decision makers after the fact in case the decision turns out badly: Managers can justify the decision by highlighting that the option with the highest expected utility was chosen after careful analysis of all the available information.[5]

The process of rationalization wastes time, effort, and money, all to uphold an appearance of having followed the culturally espoused decision process: analyses are conducted, reports are written, and presentations are given, not as a way of finding the best alternative but to justify what has already been decided.[6] Rationalization cultures are a major source of income for consulting firms. As mentioned in chapter 10, in an estimated 50 percent of cases when consulting firms advise corporations, they justify a decision already made.

The limited resources spent on rationalization could instead be applied elsewhere if organizations had decision-making cultures that value intuition and heuristics. Downstream negative consequences include cynicism and disengagement among employees who have to put in all this work, knowing full well that the decision already has been made. Moreover, when actual heuristic decision processes are disguised, learning is hindered. Without a frank discussion about which heuristics work under what circumstances, the managers' adaptive toolbox remains underdeveloped.

Rationalization cultures are more prevalent in large and bureaucratic corporations, where professional managers have little "skin in the game." Believing that more is better (i.e., more data, more analysis) and the real-world tasks that we face are small-world problems where optimization is the best way to make decisions is reinforced in business schools (see chapter 13). Making decisions in a solely analytic fashion is widely taught as a defining feature of what distinguishes professional managers.[7] Such a culture is less prevalent in family businesses and owner-led companies, which are more accepting of intuition and heuristics, so long as the results are good.[8]

CYA Culture

The defensive decision-making culture goes beyond the rationalization culture. In the latter, heuristics are used to choose a promising option for the company, but the process is rationalized post hoc. This culture allows managers to make decisions in the way that they believe is best, so long as they can rationalize the choice afterward. In a defensive culture, managers do not choose the most promising option, but instead an inferior one that can be better defended if something goes awry (as discussed in chapter 10). This culture is also known as *CYA*, which stands for "cover your ass."[9] CYA cultures exist in many organizations. For example, the purchasing department of a company was looking at two parts suppliers: a regional supplier that could provide the parts at a low price with good quality and excellent service and a global, better-known company that could provide the parts at a higher price but with lower quality and service. Nevertheless, the managers went with the better-known company, as that decision was easier to defend should any problems occur down the road.

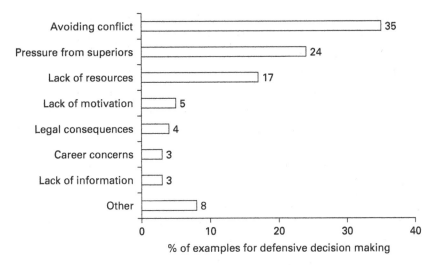

Figure 11.1

Motives for defensive decisions. Self-reported examples of 950 managers in a large public administration in Germany show avoiding conflict, pressure from superiors, and lack of resources as the most frequent reasons for defensive decisions. Based on Artinger et al. (2019).

Choosing the more recognized option is a version of the recognition heuristic. However, in this context, it is not a smart heuristic because recognition here is used not because of its correlation with accuracy but because of its correlation with defensibility: the option that is recognized can be more easily justified. Thus, heuristics are neither good nor bad in an absolute sense but only within a specific task context.

How common is defensive decision making? In a study of 950 managers in a public administration in Germany, about one quarter (25 percent) of important decisions were defensive.[10] Moreover, a large majority (80 percent) of respondents admitted to making at least one defensive decision, and 17 percent said that they made at least half of their decisions defensively.

Managers do not engage in defensive decision making because they want to, but for a number of motives that are influenced by the organization's culture. In the study of the large public administration cited above, managers' examples of defensive decision making most frequently involved avoiding conflict (figure 11.1). For instance, a manager decided not to remove a senior employee from a team to avoid conflict with him, even though the person was toxic. The second most frequent examples involved pressure from superiors. One example was a manager who offered a position to an internal candidate because their superiors preferred that, despite being convinced that an external candidate was the better one.

The motives for defensive decision making depend on the broader culture of the organization, sector, and even country. In this German public administration study, career concerns were rarely given as a reason, suggesting that these managers did not engage in defensive decision making for the selfish reason of career advancement. Concerns about legal consequences were also rarely the motive behind the defensive decisions. The situation is very different in the medical sector, where doctors and hospitals face many external pressures and are concerned about litigation: if hospitals admit mistakes, they might get sued for malpractice, especially in countries such as the US, where tort laws invite litigation. This leads to the practice of defensive medicine, characterized by overprescription and overtreatment: doctors and hospitals are less likely to get sued for carrying out an unnecessary operation or prescribing an unnecessary medicine than for not prescribing a medicine or not recommending an operation if something happens to a patient.

In a survey of US emergency physicians, 97 percent admitted ordering advanced imaging studies that they believed were medically unnecessary and explained that fear of litigation was among the primary reasons.[11] Another study found that 93 percent of US doctors practice some form of defensive decision making, including ordering clinically unnecessary magnetic resonance imaging, computed tomography, antibiotics, and surgery.[12] Just as with rationalization cultures, defensive decisions waste resources and time. In 2009, the US Congressional Budget Office estimated the cost of defensive medicine as $5.4 billion a year.[13] More recent estimates are much higher, with estimates ranging from $46 billion all the way to $300 billion, but more commonly between $50 billion and $65 billion.[14] In addition to these financial costs, defensive decisions prevent learning from mistakes and reduce quality of care, as inferior options are chosen.

Turkey Illusion Culture

The belief that an organization works in a predictable world where one knows the past, and therefore also the future, is called the *turkey illusion*. It posits a small world in which the future by definition is like the past. This creates an illusion of certainty for one's predictions. Where does the name "turkey illusion" come from?[15] Think of a turkey on the first day of its life. A human feeds it and does not kill it. The second day, the same thing happens. According to prediction models such as Bayes's rule, which assume a small world, the turkey's subjective probability that it will be fed and not killed increases day by day, and on day 100, it is higher than ever before—around 99 percent.[16] But that day is the day before Thanksgiving. The problem is that the turkey was not in a small world where everything is known, and it did not understand the underlying causality of why it was being treated well.

The turkey illusion culture is widespread in highly analytic organizations such as financial institutions. The illusion shows itself as undue confidence in the ability of quantitative models that are based on past data to predict the future. Consider the CBOE Volatility Index (VIX), which is based on the Standard & Poor's 500 index and created by the Chicago Board Options Exchange (CBOE).[17] It is also called the "fear index," and it measures the market's expectation of future volatility (Figure 11.2). Low values reflect low predicted risk. The index was at its lowest in 2007, shortly before the global financial crisis, yet at that time the actual risk was tremendously high. Or consider that financial institutions had risk models that could only predict an increase in real estate prices because, using the same reasoning as the turkey, they used data from the preceding years where prices consistently increased.

The turkey illusion also showed itself in the statements of key figures in finance. For example, as late as March 2008, Henry Paulson, the US Secretary of the Treasury, declared: "Our financial institutions, banks, and investment banks are strong. Our capital markets are resilient. They're efficient. They're flexible."[18] In 2003, Robert Lucas, a most distinguished macroeconomist, declared in his presidential address to the American Economic Association that economic theory had learned its lesson from the Great Depression and had succeeded in protecting against future disasters: "Its central problem of depression-prevention has been solved, for all practical purposes, and has in fact been solved for many decades."[19] The claim that financial crises could finally be prevented with the help of precise economic theory was reassuring—but wrong. Five years after Lucas's confident statement, the Great Recession, the worst crisis since the Great Depression, hit not only the US, but the world. The turkey was killed.

The financial crisis was caused not only by greed, wishful thinking, and poor governance. The turkey illusion also contributed to it by providing

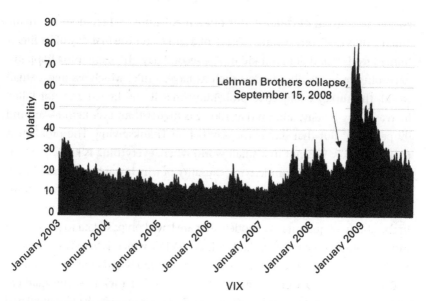

Figure 11.2

An illustration of the turkey illusion that led to the financial crisis of 2007–2008. The Volatility Index (VIX) was the first benchmark index to measure the market's expectation of future volatility, with low values indicating low expected risk. The index began decreasing consistently in 2003. In 2007, just before the global financial crisis, it was at its lowest point, and it remained low before the collapse of Lehman Brothers in 2008, suggesting illusory certainty. Data source: Chicago Board Options Exchange (http://www.cboe.com/products/vix-index-volatility/vix-options-and-futures/vix-index/vix-historical-data).

the illusion of predictability. Yet financial organizations do not operate in a small world, not even approximately.

VUCA-Denial Culture

Related to but even broader than the turkey illusion culture is the *VUCA-denial culture*. Unlike the turkey illusion culture, which mistakes a specific uncertainty for risk, VUCA-denial culture denies the existence of irreducible uncertainty itself. In such a culture, the belief exists that all VUCA (volatility, uncertainty, complexity, and ambiguity) can be tamed and reduced to small worlds. That belief allows one to calculate the optimal course of action, relying on expected utility maximization and other optimization tools. It is supported by eminent scholars in neoclassical economics, such as

Milton Friedman, who argued that all uncertainty can be reduced to risk.[20] As a result, this culture assumes that an organization could be operated like making bets in a lottery, with all options and outcomes known ahead of time. A major implication is that heuristic decision making is considered irrelevant under these assumptions.

Organizations having a VUCA-denial culture are not necessarily unaware of VUCA itself, as in the turkey illusion culture. Instead, what they deny is the irreducibility of VUCA—that it is not possible to control and reduce it completely. Thus, analysis in this culture is a tool not only of optimization but also of control, of making the unexpected expected, of trying to minimize surprises. Such a culture is common in, but by no means exclusive to, bureaucratic organizations. These organizations tend to shun uncertainty. Unpredictability gives them the shudders.

Organizations spend millions on risk management departments, regulators write hundreds of pages of regulations, and planners try to predict the future using big data, all in attempts to turn uncertainty into certainty, or at least calculable risk. Although such efforts can be successful to a certain extent, to deny VUCA and treat large worlds as if they were small worlds can result in illusions of certainty by overly detailed and thus fragile policies. Consider regulations concerning risk management in finance.[21] The first Basel Accord (Basel I) in 1988, regulating the financial industry, was 30 pages long. Basel II, passed in 2004, was 347 pages, more than ten times the earlier version. Although the regulations were meant to make the financial world safer by requiring more calculations and modeling, they did not prevent but rather contributed to the financial crisis of 2008 by creating a turkey illusion (see figure 11.2). Basel III, created in 2009 after the financial crisis, is longer still, at 616 pages. Just as in the rationalization and CYA cultures, a VUCA-denial culture leads to a misallocation of resources (of time, money, and other elements), but here, the misallocation is in trying to control the uncertainty in the financial markets by statistical tools designed for small worlds.

Manufacturing companies, especially their production departments, are among those that operate in what appears closest to a small world. The use of statistical quality control methods can lead to dramatic decreases in error rates, lower costs, and reduced waste.[22] However, the seemingly small world can be easily disrupted by unforeseen events, including global events such as pandemics, financial crises, and wars. When such events happen, the advantage of optimizing costs can turn into a disadvantage of not having kept enough slack. Optimization makes a company fragile. Such unforeseen events can disrupt global supply chains, as the severe global microchip shortage starting in 2021 has shown. Any small-world system such as a factory requires some connecting points to the larger external system for inputs such as parts and energy. Although tight control may be possible inside the smaller system, complete control is not possible over the inputs coming from the outside. As such, even in this context, VUCA cannot be controlled entirely.

How to Create a Smart Decision-Making Culture

The four cultures that we have described here are all negative. These can be replaced by three smart decision-making cultures that mutually reinforce each other. To develop these cultures, organizations need to revise three inaccurate beliefs (figure 11.3). First, organizations need to accept the reality of operating mostly in large worlds of irreducible uncertainty rather than small worlds of calculable risk. Second, they need to replace the belief that more is always better with the more nuanced—and accurate—belief that under uncertainty, less is often better. Finally, organizations need to abandon the belief that errors are always bad, in favor of the more balanced view that errors can also be useful and informative.

A Positive VUCA Culture
In response to increasing levels of VUCA, some organizations (and governments) double down on their efforts to collect even more data, to spend even more resources on risk management (and surveillance), to tighten controls even further, to increase the number of regulations, and to build ever more complex models. However, because VUCA is not fully reducible, these efforts create only illusions of certainty. The belief that a VUCA world can be tamed and controlled by big data, artificial intelligence (AI), and complex algorithms has led to pernicious consequences in the past, such as the global financial crisis.[23]

Instead, organizations need to abandon the belief that they are operating in a small world and that all large worlds can be approximated with small-world models. They need to develop a positive stance toward uncertainty. As Frank Knight observed, in a small world of risk, there is no profit.[24] Without uncertainty, there would be no innovation, no profit, and nothing new would ever happen in this world. A positive VUCA culture sees not only

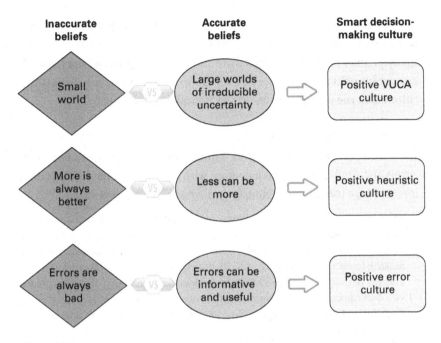

Figure 11.3
To develop smart decision-making cultures, organizations need to change their inaccurate beliefs about the world, heuristics, and errors to more accurate ones.

downsides but also upsides of operating in a large world that is not completely understood and never will be. Uncertainty offers opportunities for innovation, profit, and entrepreneurship. In such a culture, organizations focus not only on avoiding the dangers of uncertainty but also on capitalizing on its potential.

Consider the Society of Jesus, more commonly known as the Jesuits, founded in 1540 during times of great turmoil in Europe.[25] Whereas many other religious orders prescribed hundreds of regulations for their members, such as prohibitions against sleeping during lectures or wearing slippers outside the monastery, the Jesuits' founding document contained only a handful of rules. The lack of regulation enabled Jesuits to creatively take advantage of varied opportunities to carry out their missionary work. It also provided the flexibility necessary to act and adapt quickly in the uncertain environments that they encountered in their work in faraway places such as India and Japan. The absence of some rules common in other orders aimed directly at increasing flexibility, such as the lack of an obligation to pray together as a community. Some rules gave strategic direction. One such strategy heuristic was to focus their efforts on educational ventures, resulting in a legacy of many premier Jesuit educational institutions across the globe that have educated numerous leaders in society. As this example shows, a positive VUCA culture and the use of smart heuristics typically go together.

A Positive Heuristic Culture

To create a positive heuristic culture requires courage. When one of us (Gigerenzer) discussed heuristics on a podium with a member of the board of directors of a large international company, she told the following story:

> When I was new to the board, we discussed a large financial investment and all the other board members, all men, nodded in support. I didn't understand the investment but believed the others did. I didn't dare to admit my ignorance, for fear of looking foolish. The board accepted the investment and our company lost a huge amount of money. From this I learned my lesson: "Don't buy a financial product you don't understand." If someone offers me an investment, I now dare to say: "You have 15 minutes to explain how this product works. If I don't understand it, I won't buy it."

Her approach requires the courage to admit one's lack of understanding. It also requires the courage to rely on lack of understanding as a clever cue to avoid falling prey to opaque investments.

Heuristics look simple, which may be one of the reasons why managers do not admit to using them. To develop a positive heuristic culture, organizations need to endorse heuristics and intuition (the unconscious use of heuristics) as legitimate ways to make good decisions. With this book, we hope to support a move in this direction. Many family and entrepreneurial businesses already have such a culture. Entrepreneurial ventures appreciate heuristics because they are fast and frugal. This allows them to realize first-mover advantages, as well as short product cycles.[26] In many family businesses, decision making is relatively informal and intuitive, without the requirement of detailed reports including time-consuming quantitative analyses. This is especially true for owners, who do not have to justify their decisions to superiors or external parties.

For instance, Yamamotoyama, a *shinise* (Japanese for "old shop") family business, has been continuously operating in Japan since 1690. The company focuses on two products only, fine green tea and *nori* seaweed. Selecting the highest-quality ingredients is fundamental to the success of its business and is based on expert judgment rather than spreadsheets. Once a week, a small group gets together, always at the same time and at the same location, to evaluate samples. In the end, Kaichiro Yamamoto, the company president, makes the final call following his gut feeling, even if that contradicts what the company's testers say. Yamamoto told one of us (Reb) that short-term profit maximization is not the primary goal; passing the company on in a healthy state to the next generation of the family is much more important—a form of satisficing.

A Positive Error Culture

In Charles Darwin's theory of evolution, variability is the driving force. Variability is caused by errors in copying an organism's genetic material over generations. Some of these mutations are deadly, but others led to the evolution of *Homo sapiens*. Without errors, there would be no evolution—everything would remain in a small world that never changes. Similarly, in organizations, errors can have detrimental effects, but they are also indispensable for the opportunity that they afford to adapt, learn, and innovate. The challenge is to deal with errors in a way that reaps their potential benefits while containing the damage. The solution is not to reduce all errors to zero. Nevertheless, many organizations view errors in a one-sided, negative manner. The result is a dysfunctional culture.

Negative error culture: Errors are not expected to occur; if they do, one tries to hide them; if that does not succeed, one searches for a person or group to blame.

Examples of negative error cultures can be found in large corporations and hospitals, where managers are singled out for blame. Because employees are not stupid, after seeing that someone will get blamed for making an error, they hide them. Covering up errors removes the chance to talk about them and take measures to eliminate their causes. Positive error cultures are different.

Positive error culture: Errors are expected to occur; if they do, they are taken as valuable information, and one talks openly about them to identify the causes.

Positive error cultures can be found in the cockpit cultures of most commercial airlines, as well as in many family businesses. For instance, airlines have a critical incidence reporting system, and pilots go through checklist items before they take off. Crew resource management training teaches pilots and their crews to speak up when they notice a possible error, and crucially, pilots learn not to dismiss these warnings or punish copilots for speaking up.[27] As a result, flying in airplanes is very safe. If the same kind of safety training with checklists and reporting systems were established across hospitals, thousands of patients' lives could be saved every year. According to an estimate of the Institute of Medicine in 2000, between 44,000 and 98,000 lives could be saved per year in the US alone.[28] An analysis in 2013 updated this number to between 210,000 and 400,000 preventable deaths per year.[29]

Errors actually can be good, an insight that is overlooked in many education departments. For instance, most schools teach mathematics by introducing a formula and then having pupils work on various text problems that can be solved by correctly applying the formula. The goal is for pupils to make as few errors as possible. The alternative is a positive error culture, in which a problem is introduced first without the formula. The goal is to develop the ability to find solutions; to do so, errors need to be made, and pupils will learn from their errors.[30]

Not all problems have a unique solution, not even in mathematics, and in these situations, variability in opinion is not an error; it is indispensable for innovation and making progress. In other words, variability is ecologically rational in situations of uncertainty. However, variability is often confused with mere error, which needs to be reduced to zero. A prominent example of this confusion is the book *Noise: A Flaw in Human Judgment*, in which variability among judges is unconditionally identified as error.[31] To speak of error, a unique best answer would need to exist—the bull's-eye—which is often not the case in business, and certainly not in the book's two signature examples: judicial decisions and insurance underwriting.[32] Just as variability is the motor of evolution, variability in judgment is not a flaw, but the secret to success.

Error Prevention and Error Management

A positive error culture recognizes that errors can lead to both negative and positive consequences through the following two-step causal chain:[33]

Actions → Errors → Consequences

Error prevention works at the first step: to prevent actions that lead to errors. The danger is that good errors are also prevented. Many discoveries come out of errors. If an error leads to an unexpected discovery, that is an instance of serendipity. Recall 3M's invention of the Post-It Note, which we encountered in chapter 6. When trying to develop a glue, researchers in the company's research and development (R&D) department "failed," in that the glue was not strong enough. Another researcher then recognized that this weak glue could be used for an entirely different purpose: sticking small strips of paper to book pages in such a way that they adhere but also can be easily removed. Thus, an action led to an error that led to hugely positive consequences for 3M. Failure was turned into success.

Error management works at the second step. Its first goal is to prevent disastrous consequences from errors. For instance, nuclear power plants have containment shells (physical enclosures around the reactor) not to prevent errors, but to contain radiation in case a dramatic error occurs. The second goal is to reap the positive consequences of errors. For instance, the management scholar Cathy van Dyck and colleagues created a scale to measure error management culture.[34] The items are answered anonymously by employees and include the following:

- Our errors point us at what we can improve.
- When someone makes an error, (s)he shares it with others so that they don't make the same mistake.
- If people are unable to continue their work after an error, they can rely on others.

The items characterize an effective error management culture, in which errors are used to foster learning, improvement, and collaboration. In the same study, employees described many departments in their organization as having a dysfunctional error culture: "In this organization, we don't talk about errors." Another manager said: "Well, I accept errors in the sense that when a person makes too many, they're fired."

The researchers found that a score in error management culture that is one standard deviation higher (on their scale) was associated with 20 percent higher profitability. A likely reason for this finding is that a positive error culture improves decision making. For instance, the decision researcher Florian Artinger and colleagues found that a positive attitude toward errors as well as greater employee voice (i.e., the tendency of employees to speak up when they see something is wrong) was associated with less defensive decision making, which in turn helps organizations avoid wasting resources on consulting firms and other defensive rituals.[35]

To err is human and to forgive is divine. Nevertheless, the tendency to blame and punish people for their errors is deeply ingrained in management. To forgive errors should not require divinity but be part of an error culture with a human face.

12   Artificial Intelligence and Psychological Intelligence

IBM had a landmark year in 2011. On an episode of the popular quiz show *Jeopardy!* broadcast on February 16, Watson, the company's prized supercomputer, beat Ken Jennings, arguably the best human contestant, and won a $1 million prize. It was a feat that scientists and engineers at IBM had dreamed of for years, and a golden marketing opportunity for the company. Riding on the fresh publicity, IBM announced the very next day: "Already, we are exploring ways to apply Watson skills to the rich, varied language of health care, finance, law and academia." Executives at IBM were confident that the cutting-edge algorithms (mainly in natural language processing) and immense computing power behind Watson would be the engine driving the company's growth in the next decades, similar to what mainframe computers did in the previous decades. Their ambition, however, ended with a big letdown. Watson struggled to generate revenue, and the stock price of IBM in 2021 was down 10 percent from ten years before after Watson's triumph. What went wrong?

According to a report by the *New York Times*, it appears that IBM grossly underestimated the difficulty that Watson would have solving real-world problems.[1] Unlike *Jeopardy!*, in which the rules are fixed and answers to general-knowledge questions are certain, there are no clear-cut rules for cancer diagnoses, investment strategies, or research discoveries, and the outcomes are subject to many unpredictable factors. Under uncertainty, even large amounts of data, which are often messy, inconsistent, and full of errors in practice, are of limited help for building good artificial intelligence (AI) solutions. IBM claimed that Watson would be a "moon shot" that revolutionized medicine. But the claim came from their marketing department, not from their engineers, who knew better. The MD Anderson Cancer Center, for instance, spent $62 million on Watson's recommendations for cancer treatment. After the recommendations were found unreliable and some even endangered the lives of patients, they terminated the contract. IBM admitted that Watson was at the level of a first-year medical student. Soon after, Watson was sold off in parts, including the data about patients.

The failed promise of Watson is by no means an exception. The entire field of AI went through so-called AI winters—periods of reduced interest—in the 1970s and 1980s after people realized the wide gap between what they hoped AI would do and what AI could actually do. It is only with much-increased computational power, the availability of big data, and advances in machine learning that hopes for AI have been rekindled in this century. After the much-hyped victories of Watson and AlphaGo (a computerized Go program) over the best human players and aggressive marketing campaigns by tech-savvy companies such as IBM and Google, firms rushed to initiate AI-driven programs with their own enterprises. The results, however, are often underwhelming. Gartner Research reported in 2017 that 85 percent of the big data projects that it surveyed had failed to go beyond the preliminary stages,[2] and it projected in 2019 that only 20 percent of the analytic insights would actually deliver business outcomes by 2022.[3] Detailed accounts of firms' failed attempts to materialize the potentials of AI, big data, and other analytic approaches are plentiful. Given such low actual and projected return rates, it probably comes as no surprise that in a survey of 1,000 US executives in 2022, only 27 percent, 24 percent, and 11 percent of the respondents reported that their companies had used AI to "improve decision making," "improve employee experience and skill acquisition," and "enhance stockholders' trust," respectively, in the past twelve months.[4]

It would be unfair to conclude that AI has generally failed in business. AI algorithms have had a great deal of success in some domains, such as automatization and logistics. Therefore, the important question is one of ecological rationality: Under what conditions should AI be used? In this chapter, we contrast complex AI algorithms with simple heuristics, discuss them from the perspective of ecological rationality, and argue that both are needed for business to thrive in the future. The key message is that smart organizations and leaders should be aware of the limitations of complex AI algorithms and be mindful that simple heuristics can often be more helpful in decision making.

The Stable-World Principle

Why can AI algorithms beat the best humans in chess, Go, and *Jeopardy!* but are unable to outperform ordinary people in predicting recidivism and finding the right partner?[5] The answer can be derived from the distinction between small and large worlds introduced in chapter 2. The *stable-world principle* defines the domains and boundaries where AI algorithms are likely to excel.

> *Stable-world principle: Complex algorithms work better in well-defined, stable situations where large amounts of data are available. Adaptive heuristics evolved to deal with uncertainty, independent of whether big or small data are available.*

This principle makes it clear why AI algorithms deliver excellent results for some problems but not others. Watson's success on *Jeopardy!* but failure in medical research is an example, because unlike *Jeopardy!*, treatment of cancer is not a well-defined problem with stable rules.

Herbert Simon is one of the fathers of AI. AI in his work includes the analysis of the heuristics that experts use in problem solving and their incorporation in software to make computers smart. Heuristic search was part of the progress in AI and could deal with uncertainty and intractability, which the earlier, logic-based AI could not. This is why there is no real competition between AI and heuristics. The great successes of AI in chess and Go, however, are not based on this program of psychological AI, but rather on brute computational power. Recall from chapter 2 that psychological AI analyzes the heuristics that humans use and implements them in algorithms to make AI smarter. Today, most machine-learning algorithms try to solve problems without using any knowledge about the evolved brain. Although complex networks are called "deep artificial neural networks," they have little in common with human intelligence and are basically sophisticated, recursive versions of nonlinear multiple regressions. Therefore, the contrast to be made is not between AI algorithms in general and heuristics, because heuristics, such as $1/N$ and fast-and-frugal trees, are also algorithms. The contrast is between complex algorithms, such as random forest and deep learning, on the one hand, and simple, adaptive algorithms (heuristics) on the other.

The stable-world principle helps clarify the relation between complex algorithms and heuristics. If a problem is well defined and stable over time, complex algorithms and big data likely pay off; if not, simple heuristics can be equally accurate or better, while being transparent and understandable. In what follows, we give a few examples. In each, we contrast solutions derived from psychological AI—that is, simple heuristics inspired by psychology— with those produced by complex machine-learning algorithms.

Predicting Customer Purchases

In chapter 2, we mentioned the hiatus heuristic, which experienced managers use to predict whether a customer will continue making purchases. This one-clever-cue heuristic classifies a customer as inactive if the customer has not made a purchase in x months, and otherwise as active. According to an article in the *New York Times*, airlines had been using the hiatus heuristic to classify their frequent flyers since at least the 1980s.[6] Yet most research builds and refines complex models rather than trying to find out how experienced managers actually predict future purchases and learning from it.

Two marketing researchers, Markus Wübben and Florian von Wangenheim, examined the prediction accuracy of the hiatus heuristic compared to that of two widely used stochastic models, the Pareto/NBD (negative binomial distribution) and the BG/NBD (BG = beta geometric) models.[7] They tested these models in three companies, each of which provided more than 2,000 customer records. The hiatus heuristic was found to make the most accurate predictions. Interestingly, values of the heuristic's sole free parameter (i.e., the x months being dormant) that researchers estimated would give it the highest accuracy were very close to those used intuitively by managers working at the respective companies (i.e., around nine months).

A follow-up study included twenty-four more companies in the retail industry.[8] It also included two machine-learning algorithms, random forest and regularized logistic regression, as more powerful competing models. As shown in figure 12.1, the two machine-learning models did predict more accurately than the two stochastic models; their predictive accuracy, however, failed to surpass that of the hiatus heuristic. Customers' purchasing activities do not happen in a stable world; too many factors can influence the outcomes. Here, less can be more.

Inspired by these findings, a group of researchers in Berlin interviewed managers to find out how they predicted future revenue generated by customers.[9] These managers worked at a technology company that sold in-app products for mobile games (e.g., special gear and characters). They often

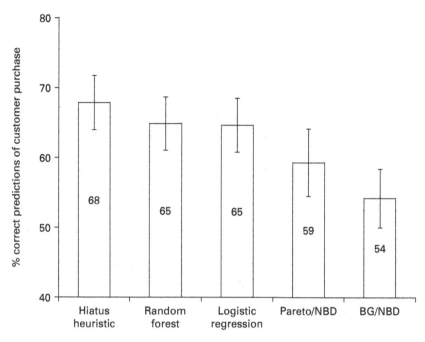

Figure 12.1
The hiatus heuristic can predict customers' purchase activities as well as or better than machine-learning algorithms (random forest and regularized logistic regression) and stochastic models (Pareto/NBD and BG/NBD). The results are based on consumer data of twenty-four retail companies. Error bars indicate standard errors. NBD = negative binomial distribution; BG = beta geometric. Based on Artinger et al. (2018).

needed to project the annual revenue generated by a customer after the customer played a game for only seven days, to help the company identify high-value customers early on. A frequently mentioned strategy was the *multiplier heuristic*: Multiply the revenue that a customer generated in the first seven days by a constant of 6. A general form of the heuristic is as follows:

> *Multiplier heuristic: Predict that the future annual sales revenue of a customer, a product, or a store is the revenue generated in an observation period multiplied by a constant X.*

The researchers then tested the forecasting accuracy of the heuristic in five mobile games. In each game, the number of customers whose purchase records were used for the test was quite large, ranging from 42,183 to 215,653. Two versions of the heuristic were examined: the original multiply-by-6, which has no free parameter, and a version that treats the multiplier as a free parameter adjusted for each game. Making the multiplier adjustable did not bring extra benefit, as the two versions had the same level of predictive accuracy. Crucially, both were as accurate as three machine-learning algorithms: LASSO (least absolute shrinkage and selection operator) regression, ridge regression, and random forest (see the upper part of figure 12.2).

To examine how general the results are, the researchers applied the multiplier heuristic to make revenue predictions beyond in-app purchases in mobile games. These included annual revenues generated by individual customers shopping in a store, by certain products made by a company (e.g., carbonated beverages of a soft-drink company), and by individual stores of a retail chain (e.g., Walmart). Altogether, they gathered fifteen data sets of such tasks, and the data points in each ranged from as few as 13 to as many as 33,520. The multiply-by-6 version of the heuristic that was developed by managers for a specific task—that is, predicting in-app purchase revenue in mobile games—no longer worked in the new tasks, as both the prediction target and domain had changed. However, when the only parameter of the heuristic, the multiplier, was estimated from data for each new task, the adjusted version of the heuristic did very well: Multiply-by-X had a lower prediction error than the three machine-learning algorithms (see the lower part of figure 12.2). This shows how heuristics can be adapted to new tasks.

Psychological AI

The multiplier heuristic and the hiatus heuristic are instances of psychological AI.[10] Psychological AI corresponds to the original vision of AI by Herbert Simon, Allen Newell, and others: to analyze how experts make decisions and to program expert heuristics into software to make computers smart. This approach is fundamentally different from most of the machine-learning approach, which relies on statistical algorithms and neglects how the human brain deals with problems in large worlds. For instance, a young child can recognize cats after seeing just one or a few; a deep artificial neural network has no such concept of a cat and needs to be trained on thousands of pictures to match the performance of children.

Generative AI such as ChatGPT is another type of deep neural network. It is special and popular because the general public can directly interact with it. ChatGPT performs amazingly well in producing answers to questions. Users tend to assume that ChatGPT "understands" their questions; yet that

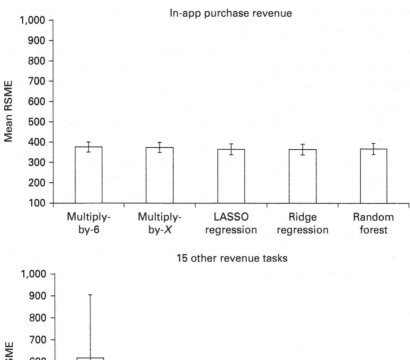

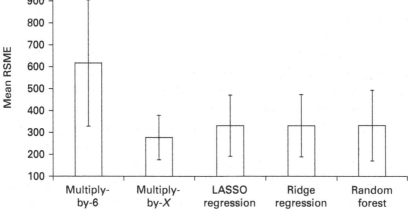

Figure 12.2

The multiplier heuristic predicts in-app purchase revenue as well as complex machine-learning algorithms, using the multiplier 6 provided by managers (top panel). For different revenue prediction tasks, the same multiplier works less well but a different multiplier can be estimated from the data (multiply-by-*X*), which results in better performance than the machine-learning algorithms that also estimate their parameters (bottom panel). RMSE = Root-mean-square error. Error bars show standard errors. These bars are much larger when performance is averaged across fifteen different tasks (bottom panel) than when it is averaged across five instances of the same in-app purchase task (top panel). Based on Artinger, Kozodi, and Runge (2020).

is not how generative AI works. As its name indicates, generative AI produces the most likely word given the previous words, just as when you type into your smartphone and a recommendation algorithm makes suggestions. The way that it generates language is fundamentally different from the way that humans do it. Large language models such as ChatGPT work by probability, not by truthfulness. The more data that it has about a topic, the more likely it is to get the answer right. We call the sentences that it gets wrong "hallucinations," but generative AI does not hallucinate; it is just a statistical prediction machine. And that is why it needs much energy. The energy consumption of GPT-3, the original release of ChatGPT, was over 1,200 megawatt hours (enough to supply an average US household for 120 years), simply for training, not counting its usage.[11] In contrast, the human brain runs on 20 watts, less than the average light bulb.

The human brain evolved to operate with little data, limited energy, and high uncertainty. AI can use this evolved wisdom.

Selecting Better Employees

Hiring good employees is critical to the development of an organization, but projecting which applicants will perform well on the job is challenging. The fit of the applicants' skills and personalities to their work team, the significant life events happening to them during their tenure, and unexpected events such as changes in leadership can all affect job performance, making a seemingly good hire at the time of recruitment a bad one some time later. Hiring is a large-world problem riddled with uncertainties.

As we saw in chapter 4, the delta-inference heuristic can help managers decide which of two job applicants to hire. Managers using the heuristic check cues sequentially and choose the applicant who is better by a threshold of delta on the first cue; otherwise, they move to the second cue, and so on. Delta-inference, like the previous heuristics, aims to describe how people make decisions and thus is another instance of psychological AI. Can it make better decisions about whom to hire than complex machine-learning algorithms can? We tested the selection accuracy of delta-inference and ordinary logistic regression in a real-world task that included over 50,000 paired comparisons formed by pairing 236 job applicants (see figure 4.4). We then examined the performance of three machine-learning algorithms: LASSO regression, random forest, and support vector machine (SVM).[12] As illustrated

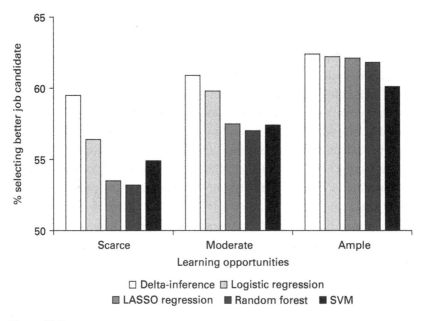

Figure 12.3

In a personnel selection task, delta-inference selected the better applicant more often than machine-learning algorithms. The advantage held regardless of whether learning opportunities were scarce, moderate, or ample (random samples of size 30, 100, and 1,000, respectively), but it was particularly pronounced when opportunities were scarce or moderate. Based on Luan et al. (2019).

in figure 12.3, delta-inference selected the better applicant more often than all other algorithms whether opportunities for learning were scarce, moderate, or ample.[13] The differences were especially pronounced when learning opportunities were scarce.

Unlike in the studies on the hiatus and multiplier heuristics discussed previously, we did not derive the parameters of the delta-inference heuristic (i.e., the cue search orders and the delta in each cue) from managers' experience. Instead, we took a data-driven approach, similar to how most machine-learning algorithms are developed. In light of the findings of this study (and many others covered in this book), we would advise data analysts to always be mindful that simple heuristics can do as well as or better than complex algorithms in situations of uncertainty, and we encourage them to try such models in their own data. The machine-learning community has echoed the same sentiment in recent years.[14]

Flagging High-Risk Loans

By the third quarter of 2022, the total value of nonperforming loans in Chinese commercial banks had amounted to 3 trillion RMB (roughly $426 billion).[15] This is just the official number—the actual number is likely higher. How can banks make better loan decisions? A collaborator of one of us (Luan) worked in one of the largest banks in China for over a decade, specializing in loans to small and medium-sized companies. She compiled data on 411 companies that were given a loan by the bank and whose payment outcomes (i.e., on time or default) were known. She identified seventeen cues commonly checked in company loan applications and codified these cues for each company. How can banks use these cues to classify loan applications as either high risk or low risk?[16]

We first recruited nineteen bank managers to establish a baseline of performance. On average, these bank managers had over ten years of experience in the loan business. We gave each manager twenty loan applications, each containing the values of the seventeen cues. We then asked them to classify the applications as either "high risk" (reject) or "low risk" (approve). Second, we constructed fast-and-frugal trees for this classification task using the four cues that were most indicative of good loan performance. For four cues arranged in the same order, eight fast-and-frugal trees can be constructed (for three cues, it is four trees; see figure 4.3 in chapter 4). These trees differ in the balance between the two possible errors: false positives (accepting an application that would later default) and false negatives (rejecting an application that would not default). In the bottom panel of figure 12.4, these eight trees are placed on the receiver-operating curve. This curve connects the individual trees that differ in how they balance the two possible errors. In general, the trees on the left side of the receiver-operating curve reduce false positives at the cost of increased false negatives, whereas those on the right side reduce false negatives at the cost of increased false positives. A glance at the curve shows that neither the leftmost tree (i.e., FFT_{HHH}) nor the four rightmost trees strike a reasonable balance between the two errors; thus, the choice of the bank should be among the three remaining trees, two of which are shown in the top panel of figure 12.4.

For instance, a bank manager using the tree shown in the top-left panel would first ask whether the credit report contains a flaw; if yes, the application is classified as high risk and rejected. If no, the second question is

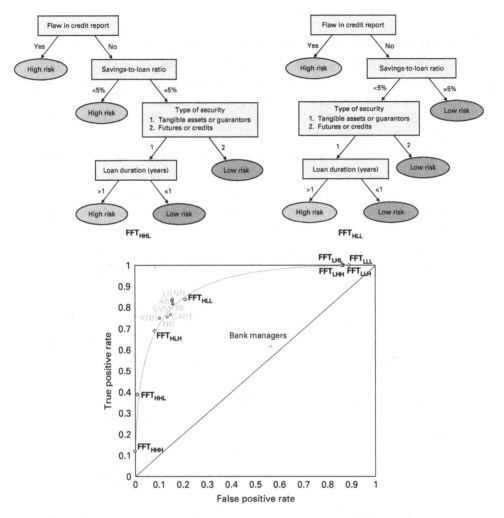

Figure 12.4

Bank managers can improve their loan decisions using fast-and-frugal trees. In addition, the transparent fast-and-frugal trees do as well as more complex and opaque machine-learning algorithms. The top panel shows two trees; the bottom panel shows the receiver-operating curve for all eight possible trees, as well as the performance of bank managers and eight machine-learning algorithms. Each fast-and-frugal tree consists of four cues arranged in the same order, and a high-risk exit flags companies that are more likely to default, thus suggesting a "reject" decision. The trees are named after the types of exits in the first three cues; an FFT_{HHL}, for instance, has three exits pointing to "high risk," "high risk," and "low risk," respectively. The true and false positive rates of bank managers were based on 380 decisions made by nineteen managers. FFT = fast and frugal tree; LR = logistic regression with L2 penalty; KNN = k-nearest neighbor; NB = naive Bayes; CART = classification and regression tree; RF = random forest; AB = adaptive boosting; NN = neural network; SVM = suport vector machine. Based on Li et al. (2022).

whether the savings-to-loan ratio is less than 5 percent; if yes, the application is classified as high risk. If no, a third question is asked, and so on. Note that the first two exits are "high risk," thus helping managers avoid false positives. In contrast, in the tree in the top-right panel, only the first exit is "high risk," thus allowing more false positives but achieving a higher true positive rate (i.e., avoiding false negatives).

The solid diagonal line in the bottom panel of figure 12.4 represents chance performance. For instance, if all loans are accepted, the false-positive and the true-positive rates are both 1, which corresponds to the point in the top-right corner. If half the loans are randomly accepted, the false-positive and the true-positive rates are both 0.5. One can see that the performance of the bank managers was only slightly above chance. They would do much better by relying on one of the fast-and-frugal trees identified in this discussion.

Can complex and opaque machine-learning techniques make better loan decisions than transparent fast-and-frugal trees? We tested eight powerful machine-learning algorithms, including SVM, random forest, and neural network. These algorithms use all seventeen cues and weight and add them in often complex ways. The bottom panel of figure 12.4 shows that on average, they do not achieve better performance than the fast-and-frugal trees. As measured by a metric called d', which balances the rates of false and true positives, the performances of the machine-learning algorithms and the fast-and-frugal trees were similar, with d' being around 1.90 for all of them.

In contrast, the performance of the bank managers was surprisingly poor, with a miserly d' of 0.13 (and chance-level accuracy being 0). Similar results of expert performance have been found in medicine. Indeed, frustration over emergency room doctors' poor performance in classifying heart attack patients was the main reason for the medical researchers Lee Green and David Mehr to develop one of the first fast-and-frugal trees in medicine.[17] One solution to this problem is to train doctors and bank managers to systematically develop and use fast-and-frugal trees. In the book *Classification in the Wild*, Konstantinos Katsikopoulos and colleagues describe how fast-and-frugal trees achieve the same feat in many other classification tasks, as well as how to construct a fast-and-frugal tree on the basis of both quantitative and qualitative data.[18] Overall, this study is another demonstration of how heuristics can be effective tools in a large world. Importantly, unlike most machine-learning algorithms, fast-and-frugal trees are totally transparent, which allows managers to understand, teach, and modify them.

Out-of-Population Prediction

In all the studies described in this chapter, the *cross-validation* method was used to test the prediction accuracy of a heuristic or algorithm. In the basic form of cross-validation, a data set is divided into two parts: the learning sample and the testing sample. A model's free parameters are estimated in the learning sample, and with these parameter values, the model is applied in the testing sample: the model's accuracy there is its predictive accuracy. This approach is also known as *out-of-sample* prediction and is generally practiced in machine learning and data analytics. It is an improvement over data fitting, in which model parameters are estimated on the entire data set and a model's accuracy is determined by how well it fits the data. Data fitting tends to "explain" patterns caused by random noise, leading to overfitting. The use of fitting instead of prediction is a problem still unrecognized by many business researchers and practitioners. It contributes to the *illusion of complexity*, the belief that complex strategies would always be more accurate than simpler ones. Complex strategies with many free parameters can fit data better, but not necessarily predict better out of sample.

Models good at out-of-sample predictions, however, can still encounter problems when they are applied to make *out-of-population* predictions, where models trained on data sets that represent one population are used to predict patterns of another population. Here, population is loosely defined. It can be a group of people, a domain of business, or events in a certain time period or location. In a VUCA world, the generalizability of a model across populations can be highly questionable. In the case of loan decisions, for instance, effective models developed for small companies operating in large cities in the 2010s may stop working for other types of companies, or even the same types of companies in the 2020s, as the cues, policies, and economic environment can all change. The multiply-by-6 heuristic provides another illustrative example: it works well for predicting in-app purchase revenue, but not different kinds of revenues. Here, the solution is to estimate the multiplier for other domains from the data. In the following discussion, we give two more examples of this issue in health care.

After the outbreak of the COVID-19 pandemic, hospitals and medical researchers around the world developed hundreds of AI algorithms to help diagnose patients and manage resources. Effective algorithms would not only save the world but also have tremendous commercial potential. But in 2021, two years into the pandemic, several review studies declared that the algorithms were largely useless, and some might even be harmful.[19] There are a multitude of reasons for this colossal failure. Not being able to generalize an algorithm beyond the data set on which it was trained is a major one.

According to Dereck Driggs, a coauthor of one of the review studies, their group at Cambridge University trained their algorithm on a data set with chest scans taken when patients were either lying down or standing up.[20] Because those who were scanned while lying down tended to be more seriously ill, the algorithm used this highly indicative but spurious cue (i.e., body position) in classifying high- and low-risk patients. In another case, researchers trained their algorithms on scans of healthy children as instances of non-COVID-19 patients. As a result, the algorithms learned to distinguish children from adults, but not the uninfected, most of whom were adults, from the infected, so they were of little diagnostic value. These examples show that even when out-of-sample prediction is excellent, out-of-population prediction can still fail because algorithms may pick up cues that are irrelevant to the task.

Epic Systems is the largest health-care software company in the US. By 2021, its software was used in over 2,400 hospitals worldwide and to manage the medical records of about two-thirds of the entire US population. Armed with this abundance of data, Epic developed various AI-powered algorithms for medical diagnoses. Its model for identifying sepsis, for example, has been used widely in US hospitals. Because the model is proprietary, as are most black-box (opaque) algorithms, few outside the company know how it works, but that does not prevent researchers from testing its diagnostic validity. In a study, a group of researchers found that among the 2,552 sepsis patients of 38,000 hospitalizations, Epic's model missed diagnosing 67 percent of them; further, among the 7,000 sepsis alerts that the model issued, only 12 percent turned out to be valid, causing a huge number of false positives.[21] Overall, using the model not only puts many patients in danger but also wastes a large amount of hospital resources.

This study is not a single case. Another study found that the accuracy of Epic's sepsis model had declined over the years and was barely above chance level at the end of the period.[22] The main culprit for this decline is *data shift*, which occurs when a population changes over time in a changing world, but an algorithm remains stationary since the time that it was trained. The specific reason for the failure of Epic's model was twofold: a change made in a new disease-coding system that was not updated in the model and an influx of a new group of patients. Having realized the problems, Epic has revised the model. But whether the new model will be much of an improvement remains to be seen.

In general, issues related to out-of-population predictions are more difficult to tackle for complex algorithms than for heuristics. The algorithms are often too opaque to allow an understanding of why and when they make mistakes, thus making them difficult to improve.

In Transparency We Trust

The Chinese government used AI-powered location-tracking apps, originally developed by the tech company Alibaba, to control the spread of the COVID-19 pandemic. Each province or large city had its own tracking app, and these apps assigned each person a color code: green (free to travel), yellow (restricted travel), or red (no movement outside residence or a confined location). One of us lived in Beijing and traveled to Shanghai in the summer of 2021. One day, his code in the Beijing app turned yellow, but the one in the Shanghai app remained green. He called all kinds of government offices to ask why and pleaded to get the yellow code revoked, as it prevented him from buying train or flight tickets back to Beijing. The answer was always "We have noted your case and will get back to you in due time." One week passed, two weeks passed, and it finally turned green in the fourth week. By then, he had missed several important in-person meetings, had a prolonged but forced stay in Shanghai, and had become very angry at the tracking apps. Countless people in China had had similar issues with these apps, and some suffered horribly.[23]

This personal story demonstrates a host of problems with decisions made by black-box algorithms. First, they are powerful but stubborn; once a decision is made, it can be very difficult to overturn. Second, human operators generally do not know how the algorithms make decisions and have little idea how to fix an error once it occurs. Third, they apparently learn slowly from feedback, perhaps because it takes a long time to locate a programming error, or because the system is so complex that changing one code may inadvertently cause other errors. Fourth, they take the responsibility of decision making away from humans, which makes it convenient for scaling but difficult to pin down who (or what) is at fault if the decision goes wrong. Last, they are subject to abuse by those in control, and even when not being abused, there is the suspicion of abuse because the decision process is so opaque to the outsiders. Black-box algorithms are not only being used in Chinese tracking apps but increasingly in organizations around the world to surveil what their employees are doing minute-by-minute and to inform decisions about whom to hire, fire, and promote.

For generations, humans have invented tools and machines to improve productivity and make life easier. Transparency has not been an important issue, as mills, cars, and phones are all assemblies of parts, and each part's functionality is known. Complex AI algorithms are a different kind of tool. The parts of which they are made are invisible to their users, and the inner working mechanisms elude most. People have an inclination to trust things that are transparent because we are able to understand them, inspect them, and improve them. So long as AI algorithms remain nontransparent, it will be difficult for people to really trust the decisions they make.

Moving Forward

In an increasingly technology-driven world, most firms are afraid of not being able to latch onto the latest trend and of being left behind. AI seems to be one such "can't-miss" trend. Before pouring millions into recruiting AI engineers, buying equipment and software, and revamping their business operations, firms should be aware of the pros and cons. As argued by the stable-world principle, complex AI algorithms can provide excellent solutions to problems that approximate small worlds and in which data are plenty and surprises are few, but they face challenges in a large world where data are often unreliable, the unexpected can happen at any time, and many factors are out of a firm's control. Moreover, black-box AI algorithms, whether effective or not, cause problems and raise concerns because of their nontransparency.

Smart heuristics work well in a large world of uncertainty. Broadly, they are also AI algorithms, but the AI here is psychological, which means two things: they are derived from human experience and intelligence, as originally intended by Simon, and they meet human psychological needs, such as transparency, trust, fairness, and privacy. This is the kind of AI that can be more helpful for a firm, its leaders, and its employees. As Tim Cook, the CEO of Apple, put it, "For artificial intelligence to be truly smart, it must respect human values. . . . If we get this wrong, the dangers are profound."[24]

Moving forward, we propose that developers should routinely investigate whether a complex, nontransparent algorithm, such as a neutral network, can be replaced with a smart heuristic that is equally accurate but transparent. To this end, the heuristics introduced in this book may provide inspiration. Moreover, governments should make it mandatory that black-box algorithms for sensitive scoring, such as credit scores, health codes, or recidivism predictions in courts, be made transparent to the public. If possible, organizations should do the same, making it clear to managers and employees what information goes into an algorithm, how it is processed, and why the decisions come out as they do.

13   What You Should Learn in Business School

One might assume that US presidents commonly have degrees from top business schools. After all, they lead the US government, an exceedingly vast, complex, and powerful organization. Yet, George W. Bush is the only president to have received a master of business administration (MBA) degree, from Harvard Business School (table 13.1).[1] And only a few other presidents even studied economics as undergraduates, among them Donald J. Trump. Coincidentally, George W. Bush and Donald Trump are currently also ranked among the worst presidents in recent US history.[2] Other presidents have degrees from a variety of fields, with law being the most frequent. For instance, Bill Clinton, Barack Obama, and Joe Biden all hold law degrees, as did Richard Nixon.

The absence of business school degrees in politics is not unique to the US. UK prime ministers studied fields such as the classics (Boris Johnson), law (Tony Blair), chemistry (Margaret Thatcher), geography (Theresa May), and metallurgy (Neville Chamberlain). None received a business degree. The situation is similar in other countries. In Germany, for example, former chancellor Angela Merkel has a PhD in physics and Helmut Kohl, the chancellor with the longest term of service, earned a PhD in history.

Business School Education—Not as Useful as You Might Think?

Why do US presidents rarely have business degrees? And why does the one who has (George W. Bush) rank poorly? The influential management scholar Henry Mintzberg sees the problem with business education as follows, focusing on the popular MBA degree:[3]

> MBA students enter the prestigious business schools smart, determined, and often aggressive. There, case studies teach them how to pronounce cleverly on situations

Table 13.1

US presidents since 1945 and their tertiary education

President	Period	Rank	Area of Study, Degree
Harry S. Truman	1945–1953	2	Law, no degree
Dwight D. Eisenhower	1953–1961	1	Military academy
John F. Kennedy	1961–1963	3	Government, BA
Lyndon B. Johnson	1963–1969	6	History, BS
Richard Nixon	1969–1974	12	Law, LLB
Gerald Ford	1974–1977	10	Law, LLB
Jimmy Carter	1977–1981	9	Engineering, BS
Ronald Reagan	1981–1989	4	Economics, sociology, BA
George H. W. Bush	1989–1993	8	Economics, BA
William J. Clinton	1993–2001	7	Law, JD
George W. Bush	**2001–2009**	**11**	**Management, MBA**
Barack Obama	2009–2017	5	Law, JD
Donald J. Trump	2017–2021	13	Economics, BA
Joe Biden	2021–?	NA	Law, JD

Shown are the areas of study and highest degrees. Only one president had an MBA (highlighted in bold; there were also none before 1945). Shown is also their 2021 C-SPAN rank (for the period after 1945) of their presidential leadership performance. BA = bachelor of arts; BS = bachelor of science; JD = doctor of jurisprudence; LLB = bachelor of laws; MBA = master of business administration.

Data sources: Miller Center, University of Virginia (https://millercenter.org/president); C-SPAN (https://www.c-span.org/presidentsurvey2021/?page=overall)

they know little about, while analytic techniques give them the impression that they can tackle any problem—no in-depth experience required. With graduation comes the confidence of having been to a proper business school, not to mention the "old boys" network that can boost them to the "top." Then what?

The answer, unfortunately and all too frequently, is poor performance, unethical behavior, and ultimately failure as leaders. Jeffrey Skilling, the infamous Enron CEO, went to Harvard Business School. He served years in federal prison for his white-collar crimes. Rajat Gupta, a former managing director of the consulting firm McKinsey, also graduated from Harvard Business School. He was convicted of insider trading and also served a jail term. An MBA degree can help graduates get to the top, but it does not necessarily prepare them to perform well—or ethically.

One might object that these are just a few bad apples. However, when Mintzberg and Joseph Lampel examined the performance of nineteen Harvard alumni who were considered the school's superstars as of 1990, they found that, post-1990, ten of them failed, in that their companies went bankrupt, they were forced out as CEO, or similarly undesirable outcomes occurred. Another four showed questionable performance. Only five of these supposed superstars did well in the long run.[4] Not exactly the track record that one might expect!

It is, of course, not only Harvard Business School graduates who frequently end up being poor leaders. The problem is more widespread and systemic. Indeed, in a study of 444 CEOs featured on the covers of business magazines such as *Forbes*, the subsequent performance of those with MBA degrees was significantly worse than the performance of those without MBA degrees. The authors attributed this finding to the CEOs' "pursuit of costly growth strategies, an inability to sustain performance, and the capacity to obtain superior private benefits in compensation." The performance gap still persisted seven years later.[5] A follow-up study by the same authors with a larger sample of 5,004 US CEOs from 2003 to 2013 replicated these results. The authors summarized their findings as follows: "We find that MBA CEOs are more apt than their non-MBA counterparts to engage in short-term strategic expedients such as positive earnings management and suppression of R&D [research and development], which in turn are followed by compromised firm market valuations."[6] Apparently, in business school education, as in decision making, less can be more, and leaders can do better without an MBA.

What about the founders of big tech companies? Did they invest their time in getting a business degree? Many did not. Steve Jobs, a cofounder of Apple, dropped out of Reed College after only one year. Jeff Bezos, the founder of Amazon, went to Princeton University to get a bachelor's degree in electrical engineering and computer science. Elon Musk, the founder of Tesla, graduated from the University of Pennsylvania having studied economics and physics. He later went to California to study at Stanford University but started a venture instead. Mark Zuckerberg, the founder of Facebook, dropped out of Harvard, where he was enrolled in computer science and psychology. Sergey Brin and Larry Page, the cofounders of Google, studied computer science at Stanford University. Apparently, for many of the founders of big tech, having a business degree was not necessary to become successful entrepreneurs and leaders. Interestingly, their successors, such as Tim Cook at Apple and Sundar Pichai at Google, often do have MBAs.

Teach the Adaptive Toolbox

Mintzberg provided an answer to why business school education frequently appears to underdeliver: it focuses too much on abstract analytic methods and case study analysis. This pedagogy may be more effective at raising confidence than at imparting effective leadership and decision-making skills. And the skills that it provides—complex quantitative analyses combined with the ability to articulate and present these convincingly—may be more useful for you to get to the top than to perform well once you get there.

We already know the problem with quantitative analyses such as expected utility maximization, decision trees, and net present values: they are designed for a world of risk, not uncertainty. In business, such small worlds include routine production processes inside a factory, as discussed in chapter 11. But as soon as logistics and supply chains beyond the factory gates are considered, control, foresight, and small-world assumptions break down. The logistics problems induced by the COVID-19 pandemic, strict lockdowns, and the global shortage of microchips starting in 2021 illustrated this clearly.

Case studies—heavily used in business schools—expect students to conduct quantitative analyses, as well as the qualitative equivalent: logical analyses and arguments. Students are exhorted to consider all the facts of a case, a strategy based on a more-is-better mindset. The goal is to present a coherent analysis and recommended plan of action. However, in the actual world of business, correspondence counts more than coherence. *Correspondence* means that the actions that organizations take match the environmental conditions, leading to positive results.[7] In other words, what ultimately matters is ecological rationality, not logical rationality. Yet business schools do not just prioritize logical rationality; they rarely even offer ecological rationality as a credible alternative. This is not to say that students do not learn anything useful in business school. They are exposed to new ideas and conceptual frameworks, engage in experiential exercises and debates, and make friends and expand their networks. However, there is room for improvement.

A Paradigm Shift toward Smart Heuristics

Business schools can, and need to, do a better job of preparing students to make effective decisions when the future is uncertain, not fully knowable, and not completely controllable. A paradigm shift toward teaching smart heuristics would help make business school education more practically useful, while being grounded in sound theoretical models and rigorous empirical research.

Unfortunately, few business schools teach the science and art of heuristic decision making. Instead, across a whole range of disciplines, from finance, strategy, and marketing to leadership and human resources, heuristics are almost invariably linked to biases, portraying decision makers negatively as rather incapable cognitive misers. The adaptive decision rules that practitioners all over the world use are delegitimized as inferior. Implicit or explicit is the message that those who use them are somehow less professional and sophisticated than their peers using complex analytic models.[8] The decision models held in highest regard are optimization and utility maximization models, which provide an air of "scientificity" that some business schools appear to desperately desire.[9]

However, as we have emphasized throughout this book, optimization is not possible in the situations of uncertainty that are common in managerial decision making, when all the future states, their outcomes, and their probabilities are not known or knowable. Thus, despite their seeming rigor and appeal, these models are of limited help in making better decisions in a VUCA world. They should not be presented as the gold standard to be applied in all contexts, cultures, and organizations. Instead, they should be taught as only one type of strategy available in the adaptive toolbox that can be useful in certain situations (i.e., under certainty or risk when the necessary information is available).

To provide a more accurate—and empowering—view of decision making, business school curricula need to change toward emphasizing ecological rationality and smart heuristics, not logical rationality and heuristics and biases. The general approach we envision is as follows:

Don't avoid heuristics—learn how to use them.

Doing so includes the following five principles:[10]

- *Take uncertainty seriously.* Teach the difference between risk and uncertainty and explain that optimization, such as expected utility maximization, is impossible under uncertainty.
- *Take heuristics seriously.* Teach the basic classes of heuristics, demonstrate their effectiveness in situations of uncertainty and intractability, and enrich managers' adaptive toolbox of strategies.

- *Analyze ecological rationality.* Match task environments with heuristics and other strategies, providing an understanding of the situations in which a particular heuristic is likely to succeed.
- *Pay attention to process.* Teach the actual process of decision making (e.g., the search, stopping, and decision rules) and the design of the external environment, and focus less on internal psychological constructs.
- *More can be less.* Teach the conditions under which complex big data models increase costs, lead to less accurate decisions, and result in less transparency.

Heuristics Are Taught Outside Business Schools

It can be informative to look beyond the confines of one's own field to learn more about the practices of others. If business schools did so, they would find that disciplines held in the highest esteem, such as mathematics and artificial intelligence (AI), actually take heuristics seriously. They portray heuristics in a positive light to their students. Two important differences from the typical business school curriculum stand out. First, in these fields, heuristics are seen as valuable, even indispensable, strategies for discovery, problem solving, and decision making under uncertainty and intractability. Second, heuristics and analysis are not portrayed in an antagonistic either–or relationship, where people use either heuristics *or* analysis. Instead, these fields recognize the value of using both, depending on the task, and of going back and forth from heuristics and intuition to analysis.

Consider mathematics. It is the most abstract of all scientific fields, and yet heuristics are greatly appreciated in mathematics and mathematical pedagogy. George Pólya's classic book *How to Solve It* is an illustration of this point.[11] In the preface to the first edition, Pólya introduced two methods of mathematics, one systematic and deductive, the other experimental and inductive. It is this second method that relies heavily on intuition and heuristics. Neither of the two methods is more important, and mathematics would be impossible without either. Rather than emphasizing that heuristics do not always lead to suitable ways of solving problems, Pólya highlighted the positive: they have the potential to do so. He went on to describe various heuristics for mathematical problem solving such as using analogy or depicting the problem in a drawing.

In the fields of computer science and AI, heuristics are also indispensable. A seminal textbook is called simply *Heuristics*. Its author, Judea Pearl, wrote, "The study of heuristics draws its inspiration from the ever-amazing observation of how much people can accomplish with that simplistic, unreliable information source known as intuition."[12] The subtitle of the book is also telling: *Intelligent Search Strategies for Computer Problem Solving*. Thus, whereas heuristics have commonly been portrayed as primitive and inferior strategies in management and parts of psychology, in computer science, they are seen as the opposite: intelligent.

Consider the famous problem in which a traveling salesperson wants to find a route between cities that minimizes the total travel distance. If the person has to visit 100 cities, a computer that can check a million routes per second would require about 2.9×10^{142} centuries to check all the possible routes to find the optimal route. Yet Earth is only 4.543×10^9, or 4.543 billion, years old.[13] Fast-and-frugal heuristics, in contrast, can find good solutions quickly. One heuristic in this case is to always proceed to the nearest city not yet visited. This *nearest-neighbor heuristic* is very easy to apply and does remarkably well.[14] Well-defined problems where no computer or mind can find the *best* solution in real time are called *computationally intractable* (see chapter 2). Intelligent search heuristics can handle these problems well.

How to Learn Heuristics

Fields such as mathematics and AI take advantage of the fact that the simplicity of heuristics makes them relatively easy to communicate, teach, learn, and apply. Business schools could easily draw on this *simplicity advantage* as well. We begin with three ways to learn heuristics: evolutionary, social, and individual learning. These are not exclusive categories; rather, they mutually reinforce each other.

Mastering Evolved Heuristics

The first way to acquire heuristics is through evolutionary learning. For instance, children imitate more precisely and generally than chimpanzees or any other species. Thus, the core capacity for imitation has evolved, but the objects of imitation (e.g., peers, parents, competitors) need to be learned. The same challenge applies in business: whom to imitate? Imitate-the-successful and imitate-the-majority are two heuristics that address this question in different circumstances. As we have seen in chapter 5, Masayoshi Son relied on his time machine heuristic, a version of imitate-the-successful, to replicate business models successful in the US in other countries. This strategy allowed Son to gain a local first-mover advantage. In contrast, KFC, an American fast-food company, used the imitate-the-majority heuristic in China and other Asian markets to adapt its menu to the tastes of local majorities. The strategy allowed KFC to increase demand by relying on proven dishes. Likewise, when a company relies on the imitate-the-majority heuristic to open up a call center in India, it can reduce uncertainty by drawing on existing infrastructure and other resources.

The gaze heuristic (discussed in chapter 3) is another example. It exploits the evolved capacity to keep one's gaze on a moving object (against noisy backgrounds), which is difficult for computers to do but easy for animals. The heuristic helps people catch flying objects without estimating their trajectories and guides missiles to hit their targets.

Although heuristics exploit evolved core capacities, this does not mean that everyone is equally proficient at applying them. Some practice is needed. The requirement of practice has implications for learning and teaching evolved heuristics, including social heuristics such as imitation. Importantly, they cannot be mastered through mere conceptual understanding. Expertise develops, more than anything, through practice. For example, business school education can integrate conceptual teaching with practice opportunities through part-time degrees in which working students can apply what they learn in the classroom right away.

To become proficient in the use of heuristics in organizations, employees need sufficient opportunities for practice. Most professional baseball players have been practicing from a young age. Also, outfielders immediately receive feedback (they caught the ball or they didn't). Such conditions can be difficult to replicate in business, especially for rare strategic decisions; other decisions are made regularly, and expertise can develop over time. However, when organizations frequently move employees from one task, job, or department to another, they reduce opportunities for repeated practice and mastery of heuristics. Although there are often good reasons for such moves, including the broadening of employees' experience, organizations need to recognize that there is often a trade-off between breadth and depth. Few professional athletes are excellent in more than one sport unless the sports draw on the same underlying capabilities.

Social Learning of Heuristics
The second major way in which heuristics are acquired is through social learning. Social learning can be explicit or implicit. Implicit learning relies

on the imitation of heuristics observed in one's social environment. Explicit learning is the outcome of being taught heuristics. The teaching of heuristics is already happening successfully in other fields, as mentioned previously, and is a prime candidate for a useful course in business schools. In addition, fast-and-frugal trees have been developed and taught in medicine for diagnosis and treatment, such as diagnosis of ischemic heart disease and treatment of catheter-associated infections.[15] Konstantinos Katsikopoulos and colleagues explain step by step how to construct fast-and-frugal trees and tallying heuristics and describe the conditions under which these heuristics can match or outperform complex machine-learning algorithms.[16] The Bank of England collaborates with the Max Planck Institute for Human Development to develop and teach simple heuristics for making the world of finance safer.[17]

Business schools can learn from these programs and imitate them. For example, constructing fast-and-frugal trees could be integrated into a human resources course on making hiring decisions—see the examples of Musk's and Bezos's hiring heuristics discussed in chapter 4. Excellent learning material already exists and could quite easily be adapted to the business context.[18]

When the goal is to learn and teach industry- and job-specific heuristics, specific knowledge may be required. For example, the organizational scholars Christopher Bingham and Kathleen Eisenhardt found in qualitative research that organizations develop portfolios of heuristics that are quite specific to their context.[19] In such cases, the first step would be to elicit the set of possible heuristics. From this, a subset of the most suitable ones would be selected. Then a training program can be developed.

This is exactly the process that the decision researcher Gavin Maistry followed when developing a simple-rules training program for insurance underwriters.[20] Underwriters decide whether and at what price to offer insurance against various risks, such as earthquakes, hacker attacks, or spaceflight. Maistry observed that while underwriters receive extensive quantitative training on assessing risks and calculating expected costs and profits, they are not trained to have good judgment. To remedy this situation, he designed a course on smart heuristics for underwriters. After interviewing professional underwriters with many years of experience to elicit these heuristics for a range of situations, he extracted a set of ten simple rules from this raw material. These rules are typically cases of one-clever-cue heuristics (see chapter 3), in which a single cue is used to make a decision on whether to accept or reject a risk. An example of a rule for underwriting is "The risk must be random and unintentional." The one-clever-cue heuristic following from the principle is "Never insure a nonrandom risk." The rationale is that if an individual deliberately (i.e., nonrandomly) causes a loss, they should not be indemnified against the loss.

In the ninety-minute program, trainees first make decisions about underwriting scenarios. Subsequently, the underwriting heuristics are revealed, explained, and discussed. In a quasi-randomized study with 220 participants, this training improved the accuracy of underwriting decisions, compared to a control condition consisting of a lecture on underwriting decision making by Daniel Kahneman (figure 13.1). The training was most beneficial for junior underwriters, defined as those with one-to-four years of underwriting experience. Midlevel and senior underwriters, defined as those with five-to-nine and at least ten years of experience, respectively, still benefited, but less so, perhaps because they had already learned most of the simple rules outside the training program. The training also improved the consistency of decisions across pairs of problems that required application of the same heuristic.

In another study, researchers conducted a randomized field experiment to compare the effects of teaching basic financial heuristics and the standard training of a typical accounting training course. The trainees were over 1,000 microentrepreneurs in the Dominican Republic.[21] The heuristics training improved financial practice, objective reporting quality, and revenues of the firms, whereas the standard training did not. Moreover, the heuristics training was particularly beneficial for microentrepreneurs with lower skills or poor initial financial practices, perhaps because their adaptive toolbox of heuristics was less developed.

Individual Learning of Heuristics

Individual learning means learning from one's own experience without the help of others (teaching) or copying the heuristics of others. It does not build directly on evolved core capacities such as imitation, but on what is called *operant conditioning* in the psychology of learning. Consider the leadership heuristics "first listen, then speak" and "hire well and let them do their jobs" described in chapter 9. The top executives who reported relying on these heuristics did not learn them in business school; rather, they developed them over time through experience and feedback. This does not

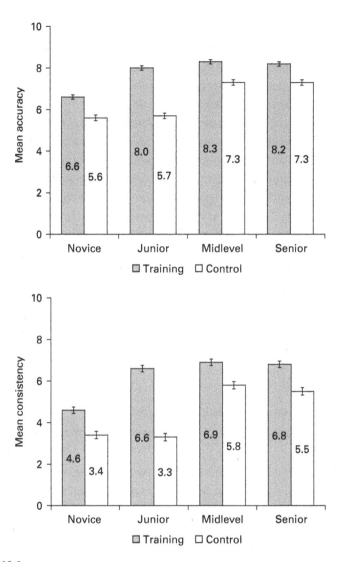

Figure 13.1
A ninety-minute systematic heuristics training led to greater accuracy and consistency of underwriting decisions compared to a control condition. The benefit was particularly pronounced for junior underwriters with one-to-four years of underwriting experience. Accuracy was measured as the number of correct decisions. Consistency was measured as the number of correct pairs of linked decisions where the same heuristic applies. The highest possible score for both is 10. The bars represent standard errors. Based on Maistry (2019).

mean that the heuristics were all unique. These top executives faced similar challenges, such as deciding whom to trust, promoting the right people into leadership positions, and making strategic decisions about the direction of the company. As a result, many of the heuristics exhibit similarities, consistent with the principle of ecological rationality.

Similarly, Bingham and Eisenhardt reported that the various heuristics learned from individual experience fall into four categories, a relatively small number.[22] These categories are selection, procedural, priority, and temporal heuristics. For example, selection heuristics are rules that guide which opportunity to pursue and which to ignore (for more detail, see chapter 5). Whereas a US company in their study used a selection heuristic to "restrict internationalization to English-speaking markets," a Finnish company instead used it to internationalize its operations, first moving into Scandinavian countries, starting with Sweden.

How can business school education be changed to support leaders and organizations in learning heuristics from their own experience? Bent Flyvbjerg has provided an instructive example.[23] He designed a workshop for teasing out leaders' heuristics through the following five sequential steps:

Step 1: Introduce the concept of heuristics, highlighting their value as decision strategies.

Step 2: Provide examples of leadership heuristics.

Step 3: Have participants select those heuristics that resonate with them and explain why.

Step 4: Have participants formulate their own heuristics that have helped them succeed, reflecting on their work experience and the heuristics discussed so far.

Step 5: Have participants share their favorite heuristics with the group and explain how the heuristics helped them.

The workshops are conducted in relatively small groups of twenty-to-thirty leaders who typically have twenty or more years of leadership experience, creating a conducive setting for reflection and open sharing. According to Flyvbjerg, a key benefit of the workshops is that leaders gain a more explicit understanding of the heuristics that they had been using intuitively and with little awareness. By making them explicit, leaders then can more easily communicate and teach them to their teams. Organizations also stand to benefit from this transfer of knowledge that was hitherto tacit and would have been lost with the departure of the leader.

1 —	Introduce smart heuristics as effective decision-making strategies (including the adaptive toolbox and ecological rationality).
2 —	Class discusses concrete examples of smart heuristics in management (e.g., why and when they are effective, what resonates).
3 —	Learners formulate their own smart heuristics explicitly (e.g., fast-and-frugal trees or one-clever-cue heuristics).
4 —	Learners share information about their best heuristics with the class (e.g., what they like about them, when they use them, how they learned them).
5 —	Learners develop one or more new heuristics drawing on their own and the group's reflection and experience (i.e., individual and social learning).
6 —	Learners try their new heuristics "in the wild," observe the results, reflect on why and when they work or not, and revise as necessary.

Figure 13.2
A six-step procedure for teaching and learning smart heuristics. The steps proceed from introducing heuristics as effective strategies (steps 1 and 2) to reflecting on and sharing smart heuristics (steps 3 and 4) to developing and applying novel heuristics (steps 5 and 6).

Inspired by Flyvbjerg, we developed a six-step process for teaching heuristics in business schools, shown in figure 13.2. Given the widespread inaccurate belief that heuristics are inferior to analysis, the first step is to describe heuristics in a positive way as smart decision strategies for a VUCA world. Next is to introduce the adaptive toolbox, ecological rationality, and different classes of smart heuristics with concrete examples in management (steps 1 and 2). Learners are then encouraged to reflect on and share their existing smart heuristics (steps 3 and 4). On this basis, they develop one or more novel smart heuristics (step 5) and use them in their decision making, carefully observing how well the heuristics do and making adjustments as necessary (step 6).

Learning How to Select Heuristics

Evolutionary, social, and individual learning create the adaptive toolbox of a manager. Managers also need to learn how to select the appropriate heuristics from their adaptive toolbox given the decision task at hand—that: is, to develop an understanding of the heuristics' ecological rationality. A number of studies have shown that individuals indeed switch between heuristics, depending on the demands of the environment. For instance,

in a situation with a dominant cue (see figure 3.6 in chapter 3), participants relied on one-reason heuristics, but when the information environment was changed to an equal-cue condition, they switched to tallying
heuristics.[24] Despite the structural changes being unknown to the decision
makers, they managed to switch to more ecologically rational heuristics by
mere individual learning from experience.

The same adaptive choice of heuristics has been shown for managers. In
an experiment, we asked seasoned managers with an average of more than
twenty-two years of supervisory experience to make decisions about hiring
or firing employees based on three cues: the employees' performance mean,
trend, and variation.[25] Using cognitive modeling, we found that the majority of managers used a fast-and-frugal tree to make this decision. Moreover,
they tended to adapt the specific structure of the tree to the decision task
(figure 13.3). Specifically, when they were asked to award a bonus to only the
top 25 percent of employees in terms of performance, the majority used a
fast-and-frugal tree with a more conservative exit structure (see figure 4.3 in
chapter 4 for an illustration of conservative and liberal fast-and-frugal trees).
When the same managers were asked to fire the worst 25 percent of performers and retain 75 percent of employees, the majority used a tree with a more
liberal structure. This adaptation of the exit structure is ecologically rational
because in the former situation, only a minority of employees could receive
a bonus, whereas in the latter, a majority of employees should be kept.

The selection of heuristics can be aided simply by updating the contents
of the adaptive toolbox. Bingham and Eisenhardt, in the study mentioned
earlier in this chapter, found that managers developed a portfolio of smart
heuristics that they regularly updated through a process of simplification
cycling: smarter heuristics replaced less effective ones, and those that were
no longer suitable were pruned.[26] This ensured that the number of heuristics
in the portfolio remained manageable, while quality increased. For instance,
a US-based enterprise software firm in their study initially had an internationalization heuristic to expand only into English-speaking markets, which
aligned with the language spoken in the organization. However, after venturing into some markets such as Australia and the UK, the organization elimi-
nated this heuristic to capture additional business opportunities in countries
such as France, Germany, and South Korea.

To encourage individual and social learning of the ecological rationality of
heuristics, organizations can expose their employees to a variety of tasks and
units, thereby allowing them to add to and refine their adaptive toolbox as

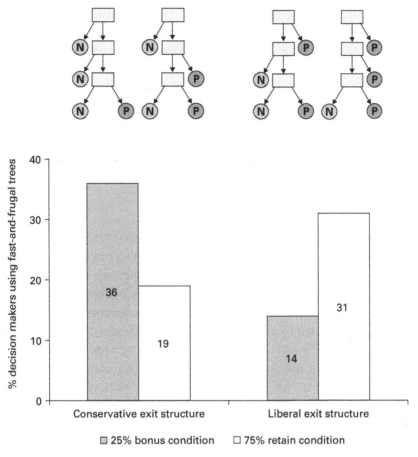

Figure 13.3
Experienced managers intuitively adapted the exit structure of their fast-and-frugal trees depending on task demands. In a condition where they had to award bonuses to only the top 25 percent of employees, the majority used one of the two trees with a conservative exit structure (e.g., grant a bonus only if all requirements are satisfied). In a condition where the same managers had to fire the worst 25 percent, the majority used one of the two trees with a liberal exit structure (e.g., keeping employees if they satisfy at least one requirement), as 75 percent of employees had to be kept. The two trees on the left are conservative, as they start with N; that is, there is a negative decision (either no bonus or not being retained) at the first branch. Meanwhile, the two trees on the right are liberal, as they start with P; that is, there is a positive decision (either bonus or being retained) at the first branch.

they learn from experience and from others. Expatriate assignments in other countries can further encourage this process by exposing leaders to the heuristics of different cultures. However, as we observed earlier in this chapter, a trade-off between breadth and depth exists. The more managers are exposed to a variety of contexts and heuristics, but for shorter periods of time, the less opportunity they have for developing mastery in their applications of these heuristics. Organizations likely benefit from a mix of generalists, who have highly varied adaptive toolboxes, and specialists, who are extremely skilled at using a more limited set of heuristics.

Rethinking the Learning of Decision Making

Worldwide, about a quarter million students are enrolled in an MBA program at any one time. They would benefit from curricula that prepare them better for the VUCA world that awaits them, in which situations are ambiguous, problems are intractable, and the future is unknown. Business school education should positively strengthen students' ability to learn, select, and apply the heuristics needed to make decisions under uncertainty. To do so, they should recognize the most fundamental distinction between risk and uncertainty. To deal with risk, the current focus on teaching analytic approaches such as probability theory, decision trees, net present value calculations, and option pricing is suitable. To deal with uncertainty, this focus needs to be extended to adaptive heuristics and their ecological rationality. This includes, among others, one-reason hiring rules, brand-name recognition heuristics, equality-based rules that generate fairness, satisficing rules for pricing, and social heuristics such as product imitation. If real-world problems contain aspects that correspond to both risk and uncertainty, a mixture of both approaches is recommended. To teach the adaptive toolbox is the first step in a curriculum; the second and greater challenge is to promote an understanding of the ecological rationality of heuristics.

At the beginning of this book, we saw how three Nobel laureates in economics approached the distinction between risk and uncertainty. At the end of this book, we provide two fundamental take-aways. First, *take uncertainty seriously; do not reduce it to risk*. Second, *do not avoid heuristics, but learn how to use them intelligently*. Taking these two principles to heart helps us make smart decisions in a world where the future is—for better or worse—full of surprises.

Glossary

1/N: A type of equality heuristic that distributes resources equally across N alternatives.

Adaptive toolbox: The repertoire of heuristics that an individual, team, or organization has acquired for making decisions.

Ambiguity: A situation where the exhaustive and mutually exclusive set of all possible future states S and their consequences C are known, but the probabilities are not; a special case of a small world.

Aspiration level: Used in the satisficing heuristic as a stopping rule to select the first alternative and end search. The level can be adaptive.

Bias–variance dilemma: The trade-off between the bias (difference between average prediction and true value) and the variance (sensitivity to irrelevant noise in the data), which constitute the total error of a prediction model. Typically, decreasing bias increases variance, and vice versa.

Building blocks: Basic components of heuristics, such as search, stopping, and decision rules; can be recombined to generate new heuristics.

Cross-validation: A procedure used to evaluate a model's predictive accuracy. Specifically, a data set is divided into two parts: the learning sample, where a model's free parameters are estimated, and the testing sample, where a model with the estimated parameters from the learning sample is applied to evaluate its predictive accuracy.

Defensive decision making: When managers rank option A as the best for the company but choose an inferior option B to protect themselves in case something goes wrong. Motivations for defensive decisions include fear of litigation and loss of reputation or employment.

Delta-inference: Search cues in the order of their validity, stop the search on the first cue where one alternative is better than the other by a threshold of delta, and choose the better alternative.

Dominant-cue condition: A situation with a powerful cue, such that the weight of the cue exceeds the sum of the weights of all other cues. If it holds, no linear model can make more accurate decisions than a one-clever-cue heuristic.

Ecological rationality: The study of the environmental conditions under which a given strategy performs better than other strategies.

Effort–accuracy trade-off: The claim that less effort, such as when relying on a simple heuristic, generally leads to less accuracy. This trade-off is true only in small worlds of risks, but not necessarily in large worlds of uncertainty.

Equality heuristics: A class of heuristics that weight all cues equally or distribute resources equally across all alternatives. Examples are tallying and $1/N$ heuristics.

Error culture: The way that an organization deals with errors. A positive error culture admits errors to learn about their causes and avoid the errors in the future. A negative error culture hides errors, and if that does not work, finds someone to blame; as a result, organizations tend to repeat the errors in the future.

False-negative rate: The proportion of negative test results among people who have a condition, such as a disease. In hiring, it refers to the proportion of good candidates who are mistakenly rejected.

False-positive rate: The proportion of positive test results among people who do not have a condition, such as a disease. In hiring, it refers to the proportion of bad candidates who are mistakenly made offers to.

Fast-and-frugal tree: A decision tree with n cues and $n+1$ exits (i.e., one exit for each cue, except two for the last cue).

Fitting: Estimate free parameters of a model based on all data in a data set and determine a model's accuracy by how well it fits the entire data set.

Fluency heuristic: Choose the first option that comes to mind. Following this heuristic is ecologically rational for experts who generate options in the order of their validity.

Gaze heuristic: Fixate your gaze on an object and adjust your running speed so that the angle of gaze remains constant. This is a heuristic used for navigation, such as for catching fly balls, landing a plane, or intercepting prey.

Heuristic: A rule of thumb that allows one to make decisions fast and with limited information searching. Heuristics are indispensable in large worlds, where uncertainty and intractability prevent one from relying on optimization tools.

Hiatus heuristic: If a customer has not made a purchase for X months or longer, classify as inactive, and otherwise as active.

Illusion of certainty: The belief that an event is absolutely certain, or under absolute control, although it is not; related to the view that all problems can be modeled by small worlds.

Imitate-the-majority: Imitate the action or practice adopted by most of your peers.

Imitate-the-successful: Imitate the most successful product, business model, or practice in a field or action taken by the most successful person.

Intractability: Refers to well-defined situations where the optimal course of actions cannot be calculated; a special case of a large world. Examples are chess, Go, and many scheduling problems.

Intuition: A feeling that (1) is based on years of experience, (2) appears quickly in one's consciousness, and (3) whose underlying rationale is unconscious. Also called *gut feeling*.

Large world: A situation that does not allow one to determine the optimal action. There are two kinds of large worlds, uncertainty and intractability. *Uncertainty* refers to ill-defined situations where the exhaustive and mutually exclusive set of all possible future states S and their consequences C are not known or knowable. *Intractability* refers to well-defined situations where the optimal course of actions cannot be calculated.

Leader's adaptive toolbox: The repertoire of heuristics that leaders have at their disposal. Skill is required for leaders to select heuristics adaptively for the problem at hand.

Less-is-more: When using less information or computation leads to more accurate decisions.

One-clever-cue heuristic: A heuristic that relies on a single reason to make a decision. Examples include the hiatus heuristic and Elon Musk's hiring heuristic (discussed in chapter 4).

One-reason decision making: A class of heuristics that base decisions on a single reason, including one-clever-cue and sequential search heuristics.

Optimization: To determine the maximum or the minimum of a function. In the context of decision making, this means finding the best course of action. Optimization is possible only in small worlds.

Out-of-population prediction: Predictions of a model in a population that is different from the population in which the model is trained.

Out-of-sample prediction: Predictions of a model that are derived following the procedure of cross-validation—that is, estimating model parameters in one part of a data set (the learning sample), and evaluating how well a model predicts in another part of the same data set (the testing sample); as opposed to fitting.

Recency heuristic: Predict that the next period's rate will be the same as the most recent rate. The heuristic is ecologically rational in fast-changing and volatile situations, such as predicting fluid market demands.

Recognition heuristic: If one of two alternatives is recognized and the other is not, then infer that the recognized alternative has a higher score on a criterion variable.

The heuristic is ecologically rational in situations where there is a strong correlation between recognition and the criterion.

Risk: A situation where the probabilities of all consequences in each possible future state are known; a special case of a small world.

Satisficing: A heuristic that sets an aspiration level α and chooses the first alternative that satisfies it. It has two versions: (1) satisficing without aspiration-level adaptation, where α is fixed; and (2) satisficing with aspiration-level adaptation, where α is changed by an amount γ after a searching period β and no satisficing alternative is found.

Small world: A situation where the exhaustive and mutually exclusive set of all possible future states S and their consequences C are known. The term is from Leonard Savage, who used the abbreviation (S, C). Unlike in large worlds, nothing new and unexpected can ever happen in a small world. If the probabilities of consequences are also known, the small world is called a situation of risk, if not, one of ambiguity.

Smart heuristic: A heuristic used in situations where it is ecologically rational—that is, expected to outperform other strategies or heuristics.

Social heuristics: A class of heuristics that relies exclusively on social input, such as word-of mouth and imitate-the-successful.

Speed–accuracy trade-off: The claim that heuristics must sacrifice accuracy for speed; true in situations of risk, but false under uncertainty.

Stable-world principle: Complex algorithms work best in well-defined, stable situations where large amounts of data are available, whereas heuristics work best when dealing with ill-defined, unstable situations of uncertainty. This principle helps one understand for what problems complex artificial intelligence algorithms are likely to succeed, and where simple algorithms or heuristics will perform better.

Take-the-best: A heuristic that searches through cues in the order of their validity and stops the search on the first cue where values of alternatives differ. Take-the-best is ecologically rational in situations where the weights of the cues decrease exponentially— that is, the weight of each cue is larger than the sum of the weights of the cues not yet searched.

Tallying: A type of equality heuristic in which a number k is set, so that if a target has k positive cue values or more, classify it as in category X; otherwise, do not. Tallying is ecologically rational if the cue weights are equal or close to each other.

Tit-for-tat: Cooperate first, and then imitate the opponent's move. It is highly effective against a broad spectrum of strategies in the iterated prisoner's dilemma game.

Transparency: A rule or algorithm is transparent if users can understand, memorize, teach, and execute it. Simple rules embody transparency.

Transparency–accuracy trade-off: The claim that algorithms, including machine learning and heuristics, must sacrifice transparency for accuracy. The trade-off does not generally hold because heuristics can be both transparent and accurate.

Uncertainty: An ill-defined situation where the exhaustive and mutually exclusive set of all possible future states S and their consequences C are not known or knowable. Unlike in a small world, uncertainty prevents optimization but is a driver of innovation. It is often confused with ambiguity, which is a small-world situation.

Unit-weighting: A type of equality heuristic that weights all cues equally in forming a judgment.

VUCA world: A world with much volatility, uncertainty, complexity, and ambiguity. In this book, it is equivalent to a large world. Specifically, V (volatility) concerns unexpected change over time, the U (uncertainty) and A (ambiguity) refer to aspects of uncertainty, and the C (complexity) to intractability.

Wisdom-of-crowds: A social heuristic in which one estimates a quantity by averaging the independent judgments of many people.

Word-of-mouth: A social heuristic in which one decides based on others' recommendations, such as asking existing employees for recommendations of suitable candidates to hire.

Notes

Chapter 1

1. Franklin (1907/1779).

2. Ariely (2008); Kahneman (2011).

3. Knight (1921).

4. Nobel Prize Outreach (2022).

5. Friedman et al. (2014, p. 3).

6. Geman, Bienenstock, and Doursat (1992).

7. Simon (1988, p. 286).

8. Kathleen Simon Frank, personal correspondence by email, January 26, 2019.

9. Bower (2011).

10. DeMiguel, Garlappi, and Uppal (2009).

11. This version is from Gigerenzer (2007).

12. Selten (1978, pp. 132–133).

13. Admati and Hellwig (2013).

14. Gigerenzer and Selten (2001).

15. For example, Gigerenzer, Hertwig, and Pachur (2011).

Chapter 2

1. Holton (1988).

2. Pólya (1945).

3. Marcus and Davis (2019).

4. Simon (1955).

5. Tversky and Kahneman (1974); Kahneman (2011).

6. Gilbert-Saad, Siedlok, and McNaughton (2018).

7. Gigerenzer, Todd, and the ABC Research Group (1999); Gigerenzer and Selten (2001).

8. Savage (1954, p. 16).

9. Savage (1954, p. 9).

10. Kay and King (2020).

11. Meda et al. (2022).

12. Luce and Raiffa (1957).

13. Tversky and Kahneman (1974).

14. DeMiguel, Garlappi, and Uppal (2009).

15. Elton, Gruber, and de Souza (2019).

16. Cited in Posner (2009, p. 287).

17. Knight (1921).

18. Kahneman (2011). There are dozens of versions of these two antithetical systems.

19. Kruglanski and Gigerenzer (2011).

20. Johnson and Raab (2003).

21. Beilock et al. (2004).

22. Klein (2018).

23. West, Acar, and Caruana (2020).

24. Baum and Wally (2003).

25. Shah and Oppenheimer (2008).

26. Artinger et al. (2018); Wübben and von Wangenheim (2008).

27. Lazer et al. (2014).

28. Katsikopoulos et al. (2022).

29. Brighton and Gigerenzer (2015); Goldstein and Gigerenzer (2002).

30. Katsikopoulos et al. (2020, p. 26).

31. Turek (n.d., pp. 7–10). See also Gunning and Aha (2019).

Chapter 3

1. Popomaronis (2021).

2. *EEOC v. Consolidated Service Systems* (1993).

3. Serwe and Frings (2006).

4. Hertwig et al. (2008).

5. Collett and Land (1975).

6. Hamlin (2017).

7. Sull and Eisenhardt (2015).

8. Luan and Reb (2017).

9. Easterbrook (2008).

10. Easterbrook (2009).

11. Luan, Schooler, and Gigerenzer (2014).

12. McCammon and Hägeli (2007).

13. Lichtman (2016).

14. Dawes and Corrigan (1974, p. 105).

15. DeMiguel, Garlappi, and Uppal (2009).

16. Hertwig, Davis, and Sulloway (2002).

17. Artinger and Gigerenzer (2016).

18. Todd and Miller (1999).

19. Tomasello (2019).

20. Galton (1907).

21. Grill-Goodman (2021).

22. Gigerenzer (2021).

23. Katsikopoulos and Martignon (2006).

24. Brighton and Gigerenzer (2015).

Chapter 4

1. Morris and Sellers (2000).

2. *NZ Herald* (2000).

3. Sackett and Lievens (2008).

4. Popomaronis (2021).

5. Schmidt and Hunter (1998).

6. Popomaronis (2020).

7. Ock and Oswald (2018).

8. Schmidt and Hunter (1998).

9. Luan, Reb, and Gigerenzer (2019).

10. Schmidt and Hunter (1998).

11. Lipshitz et al. (2001).

12. Haunschild and Miner (1997).

13. Dustmann et al. (2016).

14. Beaman and Magruder (2012).

15. *EEOC v. Consolidated Service Systems* (1993).

16. US Department of Justice (2022).

17. US Equal Employment Opportunity Commission (2022).

18. Feng et al. (2020).

19. Fifić and Gigerenzer (2014).

20. Sackett and Lievens (2008).

21. Highhouse (2008).

22. Google (n.d.).

23. See Kruglanski and Gigerenzer (2011).

24. Cappelli (2019).

25. Luan and Reb (2017).

26. Cohan (2012).

27. Blume, Baldwin, and Rubin (2009).

28. Kruglanski and Gigerenzer (2011); Melnikoff and Bargh (2018).

Chapter 5

1. SoftBank (2000, p. 4).

2. Chanchani and Rai (2016).

3. Cowan (2012).

4. Levitt (1966, p. 3).

5. McDonald and Eisenhardt (2020).

6. Golder and Tellis (1993).

7. Carlier (2022).

8. Levitt (1966, p. 4).

9. Shankar and Carpenter (2012).

10. In the terminology of the strategy researchers Pamela Haunschild and Anne Miner (1997), imitate-the-successful is called *outcome-based imitation*, and imitate-the-majority is called *frequency-based imitation*. They also distinguish a third option, *trait-based imitation*, which refers to imitating selected features of certain products.

11. Sharapov and Ross (2023).

12. Eisenhardt and Sull (2001).

13. Eisenhardt and Sull (2001).

14. See also the discussion in chapter 9 on Intel's decision making from the perspective of its leaders, Andy Grove and Gordon Moore.

15. Sull and Eisenhardt (2015).

16. Artinger and Gigerenzer (2016).

17. Thomadsen (2007).

18. Berg (2004).

19. Gigerenzer (2022a).

20. Bingham and Eisenhardt (2011).

Chapter 6

1. Cristofaro and Giannetti (2021); Gilbert-Saad, Siedlok, and McNaughton (2018); Guercini (2012); Harrison, Mason, and Smith (2015); Maxwell, Jeffrey, and Levesque (2011).

2. Manimala (1992).

3. Facebook's founder, Mark Zuckerberg, was inspired by an earlier website, Hot or Not, that allowed users to score women's and men's attractiveness on a scale of 1 to 10. Without asking for permission, Zuckerberg, then an undergraduate at Harvard University, hacked into women's dormitory websites to download pictures from their face books, which originally were paper-based directories. See Farnham (2014).

4. Hastings and Meyer (2020).

5. Schumpeter (1911, 1942).

6. Zetlin (n.d.). See also the accounts of Netflix cofounders Marc Randolph and Reed Hastings in Randolph (2019) and Hastings and Meyer (2020).

7. Sherden (1998, pp. 174–175).

8. 3M (n.d.).

9. The following is based on Lukas (2003).

10. Quote from Collins and Porras (2002, p. 150).

11. Cofounder Larry Page makes the case for 20 percent time in his 2004 *Founders' IPO Letter*, retrieved from https://abc.xyz/investor/founders-letters/2004-ipo-letter/.

12. Govindarajan and Srinivas (2013).

13. Quote from Lukas (2003). The use of male pronouns is in the original.

14. This and the next quote are from Lukas (2003).

15. Science History Institute (2020).

16. Pahl and Beitz (1996).

17. Rajshekhar (2021).

18. Vitsoe (n.d.).

19. Seifert et al. (2016).

20. Yilmaz, Seifert, and Gonzalez (2010, p. 335).

21. The following is based on Davis (2017).

22. Quote from Mangalindan (2018).

23. Stigler (1980).

24. Bondy, cited in Gigerenzer (2002, p. 23, 260).

Chapter 7

1. Gray (1998).

2. Brett (2007).

3. Caputo (2013).

4. Korobkin and Guthrie (2003, p. 798).

5. Rackham (2007).

6. United Nations (n.d.).

7. Gigerenzer (2018).

8. Rackham (2007).

9. Maddux, Mullen, and Galinsky (2008).

10. Swaab, Maddux, and Sinaceur (2011).

11. Heyes and Catmur (2022).

12. Cosmides and Tooby (1992).

13. Kumayama (1990).

14. Quote reported in Siedel (2014, p. 39).

15. Thuderoz (2017).

16. Tey et al. (2021).

17. Rapoport and Chammah (1965).

18. Axelrod (1984).

19. Nowak and Sigmund (1993).

20. Duersch, Oechssler, and Schipper (2012).

21. Oosterbeek, Sloof, and Van De Kuilen (2004).

22. Fehr and Schmidt (1999).

23. Binmore and Shaked (2010).

24. Druckman and Wagner (2016).

25. Brett (2007).

26. Galinsky and Mussweiler (2001).

27. Loschelder et al. (2016); Maaravi and Levy (2017).

28. Schweinsberg et al. (2012).

Chapter 8

1. Duhigg (2016).

2. Woolley et al. (2010).

3. Kozlowski and Ilgen (2006).

4. Salas et al. (2015).

5. Moon (2020).

6. Whitfield (2008, p. 723).

7. Brandt (2011).

8. Casali (2015).

9. Coombs and Avrunin (1977).

10. Mannes, Soll, and Larrick (2014).

11. Luan and Herzog (2022).

12. Budescu and Chen (2015).

13. Stewart (2012).

14. Tetlock (2003, p. 324).

15. Tan, Luan, and Katsikopoulos (2017).

16. Ferrazzi (2014).

17. Walther and Bunz (2005).

18. Wikipedia (n.d.).

19. Logan et al. (2010).

20. Ostrom (1990).

21. Stone (2011).

22. Deliso (2022).

23. Goodell (2011).

Chapter 9

1. The society was originally named the Kaiser Wilhelm Society and it was renamed the Max Planck Society in 1947. After Harnack, Planck served as its second president from 1930 to 1937.

2. Gigerenzer (2022b).

3. Dunbar (1998).

4. See van Vugt, Hogan, and Kaiser (2008, p. 191); see also Boehm (1999).

5. Day (2012).

6. DeRue et al. (2011). The highest correlation was .31, and many were below .10.

7. Fiedler (1964).

8. Vroom and Jago (1988, 2007).

9. Judge and Piccolo (2004).

10. Day (2012).

11. Fairchild (1930, p. 5).

12. Drucker (2006, p. 113).

13. March and Simon (1958); Simon (1947).

14. Gigerenzer (2014).

15. Maidique (2012); see also Gigerenzer (2014, pp. 115–116).

16. Walumbwa, Maidique, and Atamanik (2014).

17. Gigerenzer (2014).

18. Ma and Tsui (2015).

19. Lynn (1999, p. 164).

20. Ma and Tsui (2015).

21. Grove (1996).

22. Quigley et al. (2019).

23. *Kyodo News* (2021).

24. Flyvbjerg (2021).

25. Molinari (2020).

26. Boos et al. (2014).

27. Lord, Foti, and De Vader (1984).

28. Li, van Vugt, and Colarelli (2018).

29. See van Vugt, Johnson, et al. (2008).

30. Janson et al. (2008).

31. Lind (2001); van den Bos and Lind (2002); Proudfoot and Lind (2015).

32. Mintzberg (1973, 2013).

33. Simon (1971, pp. 40–41).

34. Davenport and Beck (2001); Goldhaber (1997).

35. McMahon and Ford (2013, p. 70).

Chapter 10

1. The Welch quote is from Akerlof and Shiller (2009, p. 14).

2. Dörfler and Eden (2019).

3. Ariely (2008).

4. Thaler and Sunstein (2008).

5. Calaprice (2011, p. 477) lists this quote as "possibly or probably by Einstein."

6. Gigerenzer (2007, 2023).

7. Gigerenzer (2014).

8. Gigerenzer (2014).

9. The following results are from Gigerenzer (2014).

10. Artinger, Artinger, and Gigerenzer (2019).

11. Klein (2018, p. 24).

12. Gigerenzer (2019).

13. Johnson and Raab (2003); Klein (2018).

14. Hertwig et al. (2008).

15. Klein (2018).

16. The German original is *"Wenn's denkst, ist's eh zu spat"* (see Eichler, 2021).

17. Beilock et al. (2004).

Chapter 11

1. Schein (1985).

2. Gigerenzer (2014).

3. Katsikopoulos et al. (2022).

4. Inland Revenue Authority of Singapore (2022).

5. Tavris and Aronson (2007).

6. Gigerenzer (2014).

7. Lejarraga and Pindard-Lejarraga (2020).

8. Gigerenzer (2014).

9. Artinger et al. (2019).

10. Artinger et al. (2019).

11. Kanzaria et al. (2015).

12. Studdert et al. (2005).

13. The estimate was given in a letter to Senator Orrin G. Hatch of Utah; https://www.cbo.gov/sites/default/files/111th-congress-2009-2010/reports/10-09-tort_reform.pdf.

14. Katz (2019).

15. The idea of the turkey illusion possibly originated in chapter 6 of the philosopher Bertrand Russell's *The Problems of Philosophy* (1912) on induction. The story was featured in Taleb and Blyth (2011).

16. According to the mathematician Pierre-Simon Laplace's rule of succession, the probability that something happens again that has happened n times so far equals $(n+1) / (n+2)$. In this case, this is about 99 percent (100/101). See Gigerenzer (2014).

17. Data for the VIX is available at https://www.cboe.com/us/indices/dashboard/vix/.

18. Quoted in Makridakis, Hogarth, and Gaba (2019, p. 796).

19. Quoted in Posner (2009, p. 287).

20. For example, see M. Friedman (2007).

21. This example appears in Haldane (2012).

22. Montgomery (2020).

23. Kay and King (2020).

24. Knight (1921).

25. The following example is from Sull and Eisenhardt (2015).

26. Eisenhardt (1989, 1990).

27. Helmreich and Merritt (2000).

28. This estimate of patients killed by preventable medical errors in US hospitals can be found in Kohn et al. (2000).

29. James (2013).

30. Gigerenzer (2014).

31. Kahneman, Sibony, and Sunstein (2021).

32. Kay (2022).

33. Keith and Frese (2011).

34. See van Dyck et al. (2005).

35. Artinger et al. (2019).

Chapter 12

1. Lohr (2021).

2. Axryd (2019).

3. White (2019).

4. Greenstein and Rao (n.d.).

5. Gigerenzer (2022a).

6. Wade (1988).

7. Wübben and von Wangenheim (2008).

8. Artinger et al. (2018).

9. Artinger, Kozodi, and Runge (2020).

10. Gigerenzer (2022a).

11. Champion (2023).

12. Luan et al. (2019).

13. Learning opportunities were manipulated through the size of a random sample (n) in which the prediction accuracy of the heuristic or a model was derived. There were three conditions of n: 30, 100, and 1,000, corresponding to scarce, moderate, and ample learning opportunities, respectively. In each condition, 5,000 samples were drawn randomly from a large database with more than 50,000 applicant pairs. See chapter 4 for more details. Figure 12.3 shows the mean prediction accuracy over these samples for each model.

14. Rudin (2019).

15. Xinhua News Agency (2022).

16. Li, Mu, and Luan (2022).

17. Green and Mehr (1997).

18. Katsikopoulos et al. (2020). One can also build FFTs using a web-based tool developed by Nathaniel Philips, Hansjörg Neth, and colleagues at https://econpsychbasel .shinyapps.io/shinyfftrees, or by running a free R package downloadable at https:// cran.r-project.org/web/packages/FFTrees/index.html.

19. Roberts et al. (2021); Wynants et al. (2020).

20. Heaven (2021).

21. Wong et al. (2021).

22. Ross (2022).

23. Dong (2022).

24. Salinas and Meredith (2018).

Chapter 13

1. Gregg (n.d.).

2. C-SPAN (n.d.).

3. Mintzberg (2017). Mintzberg's other interesting blogs on the topic of management can be accessed here: https://mintzberg.org/blog

4. Mintzberg and Lampel (2001). See also Mintzberg's book: Mintzberg (2004).

5. Miller and Xu (2016). Quote is from page 286.

6. Miller and Xu (2019). Quote is from page 285.

7. Hammond (2000).

8. Lejarraga and Pindard-Lejarraga (2020).

9. Bettis (2017); Hambrick (2007).

10. Gigerenzer et al. (2022).

11. Pólya (1945).

12. Pearl (1984, p. xi).

13. Yanofsky (2013).

14. Johnson and McGeoch (1997).

15. Naik et al. (2017); Wegwarth, Gaissmaier, and Gigerenzer (2009).

16. Katsikopoulos et al. (2020).

17. Aikman et al. (2021).

18. Gigerenzer, Hertwig, and Pachur (2011); Katsikopoulos et al. (2020); Wegwarth et al. (2009).

19. Bingham and Eisenhardt (2011).

20. Maistry (2019).

21. Drexler, Fischer, and Schoar (2014).

22. Bingham and Eisenhardt (2011).

23. Flyvbjerg (2021).

24. Pachur (2022); Rieskamp and Otto (2006).

25. Luan and Reb (2017).

26. Bingham and Eisenhardt (2011).

References

3M. (n.d.). *3M History.* https://www.3m.com/3M/en_US/company-us/about-3m/history

Admati, A., & Hellwig, M. (2013). *The bankers' new clothes.* Princeton University Press.

Aikman, D., Galesic, M., Gigerenzer, G., Kapadia, S., Katsikopoulos, K. V., Kothiyal, A., Murphy, E., & Neumann, T. (2021). Taking uncertainty seriously: Simplicity versus complexity in financial regulation. *Industrial and Corporate Change, 30,* 317–345.

Akerlof, G. A., & Shiller, R. (2009). *Animal spirits.* Princeton University Press.

Ariely, D. (2008). *Predictably irrational: The hidden forces that shape our decisions.* HarperCollins.

Artinger, F. M., Artinger, S., & Gigerenzer, G. (2019). C.Y.A.: Frequency and causes of defensive decisions in public administration. *Business Research, 12,* 9–25.

Artinger, F., & Gigerenzer, G. (2016). The cheap twin: From the ecological rationality of heuristic pricing to the aggregate market. In *Academy of Management Proceedings* (Vol. 2016, p. 13915). Academy of Management.

Artinger, F. M., Kozodi, N., & Runge, J. (2020). *Predicting revenues with the multiplier heuristic (February 28, 2020).* Available at SSRN: https://ssrn.com/abstract=3546017].

Artinger, F. M., Kozodi, N., Wangenheim, F., & Gigerenzer, G. (2018). Recency: Prediction with smart data. In J. Goldenberg, J. Laran, & A. Stephen (Eds.), *AMA Winter Academic Conference Proceedings* (pp. L-2–L-6). American Marketing Association.

Axelrod, R. (1984). *The evolution of cooperation.* Basic Books.

Axryd, S. (2019, April 16). Why 85% of big data projects fail. *Digital News Asia.* https://www.digitalnewsasia.com/insights/why-85-big-data-projects-fail

Baum, R. J., & Wally, S. (2003). Strategic decision speed and firm performance. *Strategic Management Journal, 24,* 1107–1129.

Beaman, L., & Magruder, J. (2012). Who gets the job referral? Evidence from a social networks experiment. *American Economic Review, 10,* 3574–3593.

Beilock, S. L., Bertenthal, B. I., McCoy, A. M., & Carr, T. H. (2004). Haste does not always make waste: Expertise, direction of attention, and speed versus accuracy in performing sensorimotor skills. *Psychonomic Bulletin & Review, 11,* 373–379.

Berg, N. (2004). Success from satisficing and imitation: Entrepreneurs' location choice and implications of heuristics for local economic development. *Journal of Business Research, 67,* 1700–1709.

Bettis, R. A. (2017). Organizationally intractable decision problems and the intellectual virtues of heuristics. *Journal of Management, 43,* 2620–2637.

Bingham, C. B., & Eisenhardt, K. M. (2011). Rational heuristics: The "simple rules" that strategists learn from process experience. *Strategic Management Journal, 32,* 1437–1464.

Binmore, K., & Shaked, A. (2010). Experimental economics: Where next? *Journal of Economic Behavior & Organization, 73,* 87–100.

Blume, B. D., Baldwin, T. T., & Rubin R. S. (2009). Reactions to different types of forced distribution performance evaluation systems. *Journal of Business & Psychology, 24,* 77–91.

Boehm, C. (1999). *Hierarchy in the forest.* Harvard University Press.

Boos, M., Pritz, J., Lange, S., & Belz, M. (2014). Leadership in moving human groups. *PLoS Computational Biology, 10,* Article e1003541.

Bower, B. (2011). Simple heresy: Rules of thumb challenge complex financial analyses. *Science News, 179,* 26–29.

Brandt, R. L. (2011, October 15). Birth of a salesman. *Wall Street Journal.* https://www.wsj.com/articles/SB10001424052970203914304576627102996831200

Brett, J. M. (2007). *Negotiating globally: How to negotiate deals, resolve disputes, and make decisions across cultural boundaries* (2nd ed.). Wiley.

Brighton, H., & Gigerenzer, G. (2015). The bias bias. *Journal of Business Research, 68,* 1772–1784.

Budescu, D. V., & Chen, E. (2015). Identifying expertise to extract the wisdom of crowds. *Management Science, 61,* 267–280.

Burr, I. W. (2018). *Statistical quality control methods.* Routledge.

Calaprice, A. (2011). *The ultimate quotable Einstein.* Princeton University Press.

Cappelli, P. (2019). Your approach to hiring is all wrong. *Harvard Business Review, 97,* 48–58.

Caputo, A. (2013). A literature review of cognitive biases in negotiation processes. *International Journal of Conflict Management, 24,* 374–398.

Carlier, M. (2022, November 9). *Best-selling SUV/crossover models in the U.S. 2021*. Statista. https://www.statista.com/statistics/343193/best-selling-suv-and-crossover-models-in-the-united-states

Casali, E. (2015, May 20). The table rule for the most productive team size. *Intense Minimalism*. Retrieved from https://intenseminimalism.com/2015/the-table-rule-for-the-most-productive-team-size

CBInsight. (2018, August 16). *Google strategy teardown: Google is turning itself into an AI company as it seeks to win new markets like cloud and transportation.* https://www.cbinsights.com/research/report/google-strategy-teardown

Champion, Z. (2023, April 17). Optimization could cut the carbon footprint of AI training by up to 75%. *Michigan News*. https://news.umich.edu/optimization-could-cut-the-carbon-footprint-of-ai-training-by-up-to-75

Chanchani, M., & Rai, C. (2016, December 5). Trust me, India will get my $10b: Masayoshi Son. *Economic Times*. https://telecom.economictimes.indiatimes.com/news/trust-me-india-will-get-my-10b-masayoshi-son/55803186

Cohan, P. (2012, July 13). Why stack ranking worked better at GE than Microsoft. *Forbes*. https://www.forbes.com/sites/petercohan/2012/07/13/why-stack-ranking-worked-better-at-ge-than-microsoft

Collett, T. S., & Land, M. F. (1975). Visual control of flight behaviour in the hoverfly Syritta pipiens L. *Journal of Comparative Physiology A, 99*, 1–66.

Collins, J. C., & Porras, J. I. (2002). *Built to last: Successful habits of visionary companies*. HarperBusiness.

Coombs, C. H., & Avrunin, G. S. (1977). Single-peaked functions and the theory of preference. *Psychological Review, 84*, 216–230.

Cosmides, L., & Tooby, J. (1992). Cognitive adaptations for social exchange. In J. H. Barkow, L. Cosmides, & J. Tooby (Eds.), *The adapted mind: Evolutionary psychology and the generation of culture* (pp. 163–228). Oxford University Press.

Cowan, M. (2012, February 3). Inside the clone factory: The story of the Samwer brothers and Rocket Internet. *Wired*. https://www.wired.co.uk/article/inside-the-clone-factory

Cristofaro, M., & Giannetti, F. (2021). Heuristics in entrepreneurial decisions: A review, an ecological rationality model, and a research agenda. *Scandinavian Journal of Management, 37*, Article 101170.

C-SPAN. (n.d.). *Presidential Historians Survey 2021*. https://www.c-span.org/presidentsurvey2021/?page=overall

Davenport, T., & Beck, J. C. (2001). *The attention economy: Understanding the new currency of business*. Harvard Business School Press.

Davis, J. (2017, June 4). How Lego clicked: The super brand that reinvented itself. *The Guardian*. https://www.theguardian.com/lifeandstyle/2017/jun/04/how-lego-clicked-the-super-brand-that-reinvented-itself

Dawes, R. M., & Corrigan, B. (1974). Linear models in decision making. *Psychological Bulletin, 81*, 95–106.

Day, D. V. (2012). Leadership. In S. W. J. Kozlowski (Ed.), *Oxford handbook of organizational psychology* (Vol. 1, pp. 696–729). Oxford University Press.

Deliso, M. (2022, November 12). *Kentucky city's mayoral race decided by a coin toss*. ABC News. https://abcnews.go.com/Politics/kentucky-citys-mayoral-race-decided-coin-toss/story?id=93122020

DeMiguel, V., Garlappi, L., & Uppal, R. (2009). Optimal versus naive diversification: How inefficient is the 1/N portfolio strategy? *Review of Financial Studies, 22*, 1915–1953.

DeRue, D. S., Nahrgang, J. D., Wellman, N. E. D., & Humphrey, S. E. (2011). Trait and behavioral theories of leadership: An integration and meta-analytic test of their relative validity. *Personnel Psychology, 64*, 7–52.

Dong, J. (2022, June 16). A Chinese city may have used a Covid app to block protesters, drawing an outcry. *New York Times*. https://www.nytimes.com/2022/06/16/business/china-code-protesters.html

Dörfler, V., & Eden, C. (2019). Understanding "expert" scientists: Implications for management and organization research. *Management Learning, 50*, 534–555.

Drexler, A., Fischer, G., & Schoar, A. (2014). Keeping it simple: Financial literacy and rules of thumb. *American Economic Journal: Applied Economics, 6*, 1–31.

Drucker, P. (2006). *The effective executive*. HarperCollins.

Druckman, D., & Wagner, L. M. (2016). Justice and negotiation. *Annual Review of Psychology, 67*, 387–413.

Duersch, P., Oechssler, J., & Schipper, B. C. (2012). Unbeatable imitation. *Games & Economic Behavior, 76*, 88–96.

Duhigg, C. (2016, February 25). What Google learned from its quest to build the perfect team. *New York Times*. https://www.nytimes.com/2016/02/28/magazine/what-google-learned-from-its-quest-to-build-the-perfect-team.html

Dunbar, R. I. M. (1998). *Grooming, gossip, and the evolution of language*. Harvard University Press.

Dustmann, C., Glitz, A., Schönberg, U., & Brücker, H. (2016). Referral-based job search networks. *Review of Economic Studies, 83*, 514–546.

Easterbrook, G. (2008, February 15). Time to look back on some horrible predictions. *ESPN*. http://sports.espn.go.com/espn/page2/story?page=easterbrook/080212

Easterbrook, G. (2009, February 10). TMQ's annual bad predictions review. *ESPN*. https://www.espn.com/espn/page2/story?page=easterbrook/090210

The Economist. (2021, July 15). Why Nord Stream 2 is the world's most controversial energy project. https://www.economist.com/the-economist-explains/2021/07/14/why-nord-stream-2-is-the-worlds-most-controversial-energy-project

EEOC v. Consolidated Service Systems, 989 F.2d 233. (7th Cir. 1993).

Eichler, C. (2021, August 15). Kleines dickes Müller. *Frankfurter Allgemeine Zeitung*. https://www.faz.net/aktuell/sport/fussball/bundesliga/gerd-mueller-fussball-legende-des-fc-bayern-muenchen-ist-tot-17486031.html

Eisenhardt, K. M. (1989). Making fast strategic decisions in high-velocity environments. *Academy of Management Journal, 32*, 543–576.

Eisenhardt, K. M. (1990). Speed and strategic choice: How managers accelerate decision making. *California Management Review, 32*, 39–54.

Eisenhardt, K. M., & Sull, D. N. (2001). Strategy as simple rules. *Harvard Business Review, 79*, 106–119.

Elton, E. J., Gruber, M. J., & de Souza, A. (2019). Passive mutual funds and ETFs: Performance and comparison. *Journal of Banking & Finance, 106*, 265–275.

Fairchild, D. G. (1930). *Exploring for plants*. Macmillan.

Farnham, A. (2014, June 2). *Hot or Not's co-founders: Where are they now?* ABC News. https://abcnews.go.com/Business/founders-hot-today/story?id=23901082

Fehr, E., & Schmidt, K. (1999). A theory of fairness, competition and cooperation. *Quarterly Journal of Economics, 114*, 817–868.

Feng, Z., Liu, Y., Wang, Z., & Savani, K. (2020). Let's choose one of each: Using the partition dependence effect to increase diversity in organizations. *Organizational Behavior and Human Decision Processes, 158*, 11–26.

Ferrazzi, K. (2014). Getting virtual teams right. *Harvard Business Review, 92*, 120–123.

Fiedler, F. E. (1964). A contingency model of leadership effectiveness. *Advances in Experimental Social Psychology, 1*, 149–190.

Fifić, M., & Gigerenzer. G. (2014). Are two interviewers better than one? *Journal of Business Research, 67*, 1771–1779.

Flyvbjerg, B. (2021). *Heuristics for masterbuilders: Fast and frugal ways to become a better project leader*. [Working paper]. Saïd Business School, University of Oxford.

Franklin, B. (1907). Letter to Jonathan Williams (Passy, April 8, 1779). In A. H. Smyth (Ed.), *The writings of Benjamin Franklin* (Vol. 7, pp. 281–282). Macmillan.

Friedman, D., Isaac, R. M., James, D., & Sunder, S. (2014). *Risky curves: On the empirical failure of expected utility*. Routledge.

Friedman, M. (2007). *Price theory*. Transaction Publishers.

Galinsky, A. D., & Mussweiler, T. (2001). First offers as anchors: The role of perspective-taking and negotiator focus. *Journal of Personality and Social Psychology 81*, 657–669.

Galton, F. (1907). Vox populi. *Nature, 75*, 450–451.

Geman, S., Bienenstock, E., & Doursat, R. (1992). Neural networks and the bias/variance dilemma. *Neural Computation, 4*, 1–58.

Gigerenzer, G. (2002). *Reckoning with risk: Learning to live with uncertainty*. Penguin.

Gigerenzer, G. (2007). *Gut feeling: The intelligence of the unconscious*. Penguin.

Gigerenzer, G. (2014). *Risk savvy: How to make good decisions*. Viking.

Gigerenzer, G. (2018). The bias bias in behavioral economics. *Review of Behavioral Economics, 5*, 303–336.

Gigerenzer, G. (2019). Expert intuition is not rational choice [Review of the book *Sources of power* (20th anniversary ed.), by Gary Klein]. *American Psychologist, 132*, 475–480.

Gigerenzer, G. (2021). What is bounded rationality? In R. Viale (Ed.), *Routledge handbook of bounded rationality* (pp. 55–69). Routledge.

Gigerenzer, G. (2022a). *How to stay smart in a smart world: Why human intelligence still beats algorithms*. Penguin UK.

Gigerenzer, G. (2022b). Simple heuristics to run a research group. *PsyCh Journal, 11*, 275–280.

Gigerenzer, G. (2023). *The intelligence of intuition*. Cambridge University Press.

Gigerenzer, G., Hertwig, R. E., & Pachur, T. E. (2011). *Heuristics: The foundations of adaptive behavior*. Oxford University Press.

Gigerenzer, G., Reb, J., & Luan, S. (2022). Smart heuristics for individuals, teams, and organizations. *Annual Review of Organizational Psychology and Organizational Behavior, 9*, 171–198.

Gigerenzer, G., & Selten, R. (Eds.). (2001). *Bounded rationality: The adaptive toolbox*. MIT Press.

Gigerenzer, G., Todd, P. M., & the ABC Research Group. (1999). *Simple heuristics that make us smart*. Oxford University Press.

Gilbert-Saad, A., Siedlok, F., & McNaughton, R. B. (2018). Decision and design heuristics in the context of entrepreneurial uncertainties. *Journal of Business Venturing Insights, 9*, 75–80.

Golder, P., & Tellis, G. (1993). Pioneering advantage: Marketing logic or marketing legend, *Journal of Marketing Research, 30,* 158–170.

Goldhaber, M. H. (1997). The attention economy and the Net. *First Monday. 2*(4). https://firstmonday.org/ojs/index.php/fm/article/download/519/440

Goldstein, D. G., & Gigerenzer, G. (2002). Models of ecological rationality: The recognition heuristic. *Psychological Review, 109,* 75–90.

Goodell, J. (2011, January 17). Steve Jobs in 1994: The *Rolling Stone* interview. *Rolling Stone.* https://www.rollingstone.com/culture/culture-news/steve-jobs-in-1994-the -rolling-stone-interview-231132

Google. (n.d.). Unbiasing. *re:Work.* https://rework.withgoogle.com/subjects/unbiasing/

Govindarajan, V., & Srinivas, S. (2013). The innovation mindset in action: 3M Corporation. *Harvard Business Review, 91.* https://hbr.org/2013/08/the-innovation-mindset -in-acti-3

Gray, F. G. (Director). (1998). *The negotiator* [Film]. Warner Bros.

Green, L., & Mehr, D. R. (1997). What alters physicians' decisions to admit to the coronary care unit? *Journal of Family Practice, 45,* 219–226.

Greenstein, B, & Rao, A. (n.d.). *PwC 2022 AI Business Survey.* PricewaterhouseCoopers. https://www.pwc.com/us/en/tech-effect/ai-analytics/ai-business-survey.html

Gregg, G. L., II. (n.d.). *George W. Bush: Life before the presidency.* Miller Center of Public Affairs, University of Virginia, Charlottesville. https://millercenter.org/president/gw bush/life-before-the-presidency

Grill-Goodman, J. (2021, November 16). *Report: 30% of online customer reviews deemed fake.* RIS. https://risnews.com/report-30-online-customer-reviews-deemed-fake

Grove, A. S. (1996). *Only the paranoid survive.* Doubleday.

Guercini, S. (2012). New approaches to heuristic processes and entrepreneurial cognition of the market. *Journal of Research in Marketing and Entrepreneurship, 14,* 199–213.

Gunning, D., & Aha, D. W. (2019). DARPA's explainable artificial intelligence program. *AI Magazine, 40,* 44–58.

Haldane, A. G. (2012, August 31). *The dog and the frisbee* [Speech]. https://www .bankofengland.co.uk/-/media/boe/files/paper/2012/the-dog-and-the-frisbee.pdf

Hamlin, R. P. (2017). "The gaze heuristic": Biography of an adaptively rational decision process. *Topics in Cognitive Science, 9,* 264–288.

Hammond, K. R. (2000). Coherence and correspondence theories in judgment and decision making. In T. Connolly & H. R. Arkes (Eds.), *Judgment and decision making: An interdisciplinary reader* (2nd ed., pp. 53–65). Cambridge University Press.

Hambrick, D. C. (2007). The field of management's devotion to theory: Too much of a good thing? *Academy of Management Journal, 50*, 1346–1352.

Harrison, R. T., Mason, C., & Smith, D. (2015). Heuristics, learning and the business angel investment decision-making process. *Entrepreneurship and Regional Development, 27*, 527–554.

Hastings, R., & Meyer, E. (2020). *No rules rules: Netflix and the culture of reinvention.* Penguin.

Haunschild, P. R., & Miner, A. S. (1997). Modes of interorganizational imitation: The effects of outcome salience and uncertainty. *Administrative Science Quarterly, 42*, 472–500.

Heaven, W. D. (2021, July 30). Hundreds of AI tools have been built to catch Covid. None of them helped. *MIT Technology Review*. https://www.technologyreview.com /2021/07/30/1030329/machine-learning-ai-failed-covid-hospital-diagnosis-pandemic/

Helmreich, R. L., & Merritt, A. C. (2000). Safety and error management: The role of crew resource management. In B. J. Hayward & A. R. Lowe (Eds.), *Aviation resource management* (pp. 107–119). Ashgate.

Hertwig, R., Davis, J. N., & Sulloway, F. J. (2002). Parental investment: How an equity motive can produce inequality. *Psychological Bulletin, 128*, 728–745.

Hertwig, R., Herzog, S. M., Schooler, L. J., & Reimer, T. (2008). Fluency heuristic: A model of how the mind exploits a by-product of information retrieval. *Journal of Experimental Psychology: Learning, Memory, and Cognition, 34*, 1191–1206.

Heyes, C., & Catmur, C. (2022). What happened to mirror neurons? *Perspectives on Psychological Science, 17*, 153–168.

Highhouse, S. (2008). Stubborn reliance on intuition and subjectivity in employee selection. *Industrial and Organizational Psychology, 1*, 333–342.

Holton, G. J. (1988). *Thematic origins of scientific thought: Kepler to Einstein.* Harvard University Press.

Inland Revenue Authority of Singapore. (2022). *Budget 2022—Overview of tax changes.* https://www.iras.gov.sg/docs/default-source/budget-2022/budget-2022---overview -of-tax-changes69937d71-ba59-4b39-b1e5-7e85b2504e1c.pdf?sfvrsn=8339ba5a_5

James, J. T. (2013). A new, evidence-based estimate of patent harms associated with hospital care. *Journal of Patient Safety, 9*, 122–128.

Janson, A., Levy, L., Sitkin, S. B., & Lind, E. A. (2008). Fairness and other leadership heuristics: A four-nation study. *European Journal of Work & Organizational Psychology, 17*, 251–272.

Johnson, D. S., & McGeoch, L. A. (1997). The traveling salesman problem: A case study in local optimization. *Local Search in Combinatorial Optimization, 1*, 215–310.

Johnson, J. G., & Raab, M. (2003). Take the first: Option-generation and resulting choices. *Organizational Behavior and Human Decision Processes, 91*, 215–229.

Judge, T. A., & Piccolo, R. F. (2004). Transformational and transactional leadership: A meta-analytic test of their relative validity. *Journal of Applied Psychology, 89*, 755–768.

Kahneman, D. (2011). *Thinking, fast and slow*. Farrar, Straus & Giroux.

Kahneman, D., Sibony, O., & Sunstein, C. R. (2021). *Noise: A flaw in human judgment*. Little, Brown.

Kanzaria, H. K., Hoffman, J. R., Probst, M. A., Caloyeras, J. P., Berry, S. H., & Brook, R. H. (2015). Emergency physician perceptions of medically unnecessary advanced diagnostic imaging. *Academic Emergency Medicine, 22*, 390–398.

Katsikopoulos, K. V., & Martignon, L. (2006). Naive heuristics for paired comparisons: Some results on their relative accuracy. *Journal of Mathematical Psychology, 50*, 488–494.

Katsikopoulos, K. V., Simsek, Ö., Buckmann, M., & Gigerenzer, G. (2020). *Classification in the wild*. MIT Press.

Katsikopoulos, K. V., Şimşek, Ö., Buckmann, M., & Gigerenzer, G. (2022). Transparent modeling of influenza incidence: Big data or a single data point from psychological theory? *International Journal of Forecasting, 38*, 613–619.

Katz, E. D. (2019). Defensive medicine: A case and review of its status and possible solutions. *Clinical Practice and Cases in Emergency Medicine, 3*, 329–332.

Kay, J. (2022). Daniel Kahneman, Olivier Sibony and Cass R. Sunstein: Noise: A flaw in human judgment. *Business Economics, 57*, 86–88.

Kay, J., & King, M. (2020). *Radical uncertainty. Decision-making for an unknowable future*. Bridge Street Press.

Keith, N., & Frese, M. (2011). Enhancing firm performance and innovativeness through error management culture. *Handbook of Organizational Culture and Climate, 9*, 137–157.

Klein, G. (2018). *Sources of power*. MIT Press.

Knight, F. H. (1921). *Risk, uncertainty and profit*. Houghton Mifflin.

Kohn, L. T., Corrigan, J. M., & Donaldson, M. S. (Eds.). (2000). *To err is human: Building a safer health system*. National Academies Press.

Korobkin, R., & Guthrie, C. (2003). Heuristics and biases at the bargaining table. *Marquette Law Review, 87*, 795–808.

Kozlowski, S. W., & Ilgen, D. R. (2006). Enhancing the effectiveness of work groups and teams. *Psychological Science in the Public Interest, 7*, 77–124.

Kruglanski, A., & Gigerenzer, G. (2011). Intuitive and deliberate judgments are based on common principles. *Psychological Review, 118*, 97–109.

Kumayama, A. (1990). Understanding gift giving in Japan. *The International Executive, 31*, 19–21.

Kyodo News. (2021, December 20). *Tokyo Games costs down at 1.45 tril. yen, no extra funds needed.* https://english.kyodonews.net/news/2021/12/9743088378ad-tokyo-games-costs-down-at-145-tril-yen-no-extra-funds-needed.html

Lazer, D., Kennedy, R., King, G., & Vespignani, A. (2014). The parable of Google Flu: Traps in big data analysis. *Science, 343*, 1203–1205.

Lejarraga, J., & Pindard-Lejarraga, M. (2020). Bounded rationality: Cognitive limitations or adaptation to the environment? The implications of ecological rationality for management learning. *Academy of Management Learning & Education, 193*, 289–306.

Levitt, T. (1966). Innovative imitation. *Harvard Business Review, 44*, 63–70.

Li, N. P., van Vugt, M., & Colarelli, S. M. (2018). The evolutionary mismatch hypothesis: Implications for psychological science. *Current Directions in Psychological Science, 27*, 38–44.

Li, Y., Mu, J., & Luan, S. (2022). *Improving quality of bank loan decisions* [Working paper]. Institute of Psychology, Chinese Academy of Sciences.

Lichtman, A. J. (2016). *Predicting the next president: The keys to the White House.* Bowman and Littlefield.

Lind, E. A. (2001). Fairness heuristic theory: Justice judgments as pivotal cognitions in organizational relations. *Advances in Organizational Justice, 56*, 56–88.

Lipshitz, R., Klein, G., Orasanu, J., & Salas, E. (2001). Taking stock of naturalistic decision making. *Journal of Behavioral Decision Making, 14*, 331–352.

Logan, D. W., Sandal, M., Gardner, P. P., Manske, M., & Bateman, A. (2010). Ten simple rules for editing Wikipedia. *PLoS Computational Biology, 6*, Article e1000941.

Lohr, S. (2021, July 16). What ever happened to IBM's Watson? *New York Times.* https://www.nytimes.com/2021/07/16/technology/what-happened-ibm-watson.html

Lord, R. G., Foti, R. J., & De Vader, C. L. (1984). A test of leadership categorization theory: Internal structure, information processing, and leadership perceptions. *Organizational Behavior and Human Performance, 34*, 343–378.

Loschelder, D. D., Trötschel, R., Swaab, R. I., Friese, M., & Galinsky, A. D. (2016). The information-anchoring model of first offers: When moving first helps versus hurts negotiators. *Journal of Applied Psychology, 101*, 995–1012.

Luan, S., & Herzog, S. (2022). *Around nine: The wisdom of small crowds* [Working paper]. Institute of Psychology, Chinese Academy of Sciences.

Luan, S., & Reb, J. (2017). Fast-and-frugal trees as noncompensatory models of performance-based personnel decisions. *Organizational Behavior and Human Decision Processes, 141*, 29–42.

Luan, S., Reb, J., & Gigerenzer, G. (2019). Ecological rationality: Fast-and-frugal heuristics for managerial decision making under uncertainty. *Academy of Management Journal, 62*, 1735–1759.

Luan, S., Schooler, L. J., & Gigerenzer, G. (2014). From perception to preference and on to inference: An approach–avoidance analysis of thresholds. *Psychological Review, 121*, 501–525.

Luce, R. D., & Raiffa, H. (1957). *Games and decisions*. Wiley.

Lukas, P. (2003, April 1). *3M: A mining company built on a mistake stuck it out until a young man came along with ideas about how to tape those blunders together as innovations— leading to decades of growth*. CNN Money. https://money.cnn.com/magazines/fsb/fsb _archive/2003/04/01/341016/

Lynn, R. J. (1999). *A new translation of the Tao-Te Ching of Laozi as interpreted by Wang Bi*. Columbia University Press.

Ma, L., & Tsui, A. S. (2015). Traditional Chinese philosophies and contemporary leadership. *Leadership Quarterly, 26*, 13–24.

Maaravi, Y., & Levy, A. (2017). When your anchor sinks your boat: Information asymmetry in distributive negotiations and the disadvantage of making the first offer. *Judgment and Decision Making, 12*, 420–429.

Maddux, W. W., Mullen, E., & Galinsky, A. D. (2008). Chameleons bake bigger pies and take bigger pieces: Strategic behavioral mimicry facilitates negotiation outcomes. *Journal of Experimental Social Psychology, 44*, 461–468.

Maidique, M. A. (2012). *The leader's toolbox*. https://lead.fiu.edu/resources/news /archives/the-leaders-toolbox-by-dr-modesto-maidique.html

Maistry, G. (2019). *An examination of the effectiveness of a training programme to improve decision making in insurance risk underwriting* [Unpublished doctoral dissertation]. Singapore Management University.

Makridakis, S., Hogarth, R. M., & Gaba, A. (2009). Forecasting and uncertainty in the economic and business world. *International Journal of Forecasting, 25*, 794–812.

Mangalindan, J. P. (2018, September 18). *How Jeff Bezos got the idea to sell "everything" on Amazon*. Yahoo Finance. https://uk.finance.yahoo.com/news/jeff-bezos-got-idea -sell-everything-amazon-134358788.html

Manimala, M. J. (1992). Entrepreneurial heuristics: A comparison between high PL (pioneering-innovative) and low PI ventures. *Journal of Business Venturing, 7*, 477–504.

Mannes, A. E., Soll, J. B., & Larrick, R. P. (2014). The wisdom of select crowds. *Journal of Personality and Social Psychology, 107*, 276–299.

March, J. G., & Simon, H. A. (1958). *Organizations*. Wiley.

Marcus, G., & Davis, E. (2019). *Rebooting AI: Building artificial intelligence we can trust*. Pantheon.

Maxwell, A. L., Jeffrey, S. A., & Levesque, M. (2011). Business angel early stage decision making. *Journal of Business Venturing, 26*, 212–225.

McCammon, I., & Hägeli, P. (2007). An evaluation of rule-based decision tools for travel in avalanche terrain. *Cold Regions Science and Technology, 47*, 193–206.

McDonald, R., & Eisenhardt, K. M. (2020). The new-market conundrum. *Harvard Business Review, 98*, 74–83.

McMahon, S. R., & Ford, C. M. (2013). Heuristic transfer in the relationship between leadership and employee creativity. *Journal of Leadership & Organizational Studies, 20*, 69–83.

Meda, N., Menti, G. M., Megighian, A., & Zordan, M. A. (2022). A heuristic underlies the search for relief in Drosophila melanogaster. *Annals of the New York Academy of Sciences, 1510*, 158–166.

Melnikoff, D. E., & Bargh, J. A. (2018). The mythical number two. *Trends in Cognitive Sciences, 22*, 280–293.

Miller, D., & Xu, X. (2016). A fleeting glory: Self-serving behavior among celebrated MBA CEOs. *Journal of Management Inquiry, 25*, 286–300.

Miller, D., & Xu, X. (2019). MBA CEOs, short-term management and performance. *Journal of Business Ethics, 154*, 285–300.

Mintzberg, H. (1973). *The nature of managerial work*. Harper and Row.

Mintzberg, H. (2004). *Managers, not MBAs: A hard look at the soft practice of managing and management development*. Berrett-Koehler.

Mintzberg, H. (2013). *Simply managing: What managers do—and can do better*. Berrett-Koehler Publishers.

Mintzberg, H. (2017, February 22). MBAs as CEOs: Some troubling evidence. *Henry Mintzberg*. https://mintzberg.org/blog/mbas-as-ceos

Mintzberg, H., & Lampel, J. (2001, February 19). Do MBAs make better CEOs? *Fortune, 19*, 105.

Molinari, M. C. (2020). How the Republic of Venice chose its Doge: Lot-based elections and supermajority rule. *Constitutional Political Economy, 31*, 169–187.

Montgomery, D. C. (2020). *Introduction to statistical quality control*. John Wiley & Sons.

Moon, L. (2020, October 13). Manage your team better with the simple "Rule of Five." *Trello*. https://blog.trello.com/manage-teams-with-5-things

Moorman, R. H., & Grover, S. (2009). Why does leader integrity matter to followers? An uncertainty management-based explanation. *International Journal of Leadership Studies, 5*, 102–114.

Morris, B., & Sellers, P. (2000, January 10). *What really happened at Coke*. CNN Money. https://money.cnn.com/magazines/fortune/fortune_archive/2000/01/10/271736/index.htm

Naik, A. D., Skelton, F., Amspoker, A. B., Glasgow, R. A., & Trautner, B. W. (2017). A fast and frugal algorithm to strengthen diagnosis and treatment decisions for catheter-associated bacteriuria. *PLOS One, 12*, Article e0174415.

Nobel Prize Outreach. (2022). Herbert A. Simon—facts. *The Nobel Prize*. https://www.nobelprize.org/prizes/economic-sciences/1978/simon/facts/

Nowak, M., & Sigmund, K. (1993). A strategy of win-stay, lose-shift that outperforms tit-for-tat in the prisoner's dilemma game. *Nature, 364*, 56–58.

NZ Herald. (2000, June 30). *Coca-Cola pays fortune to fallen boss*. https://www.nzherald.co.nz/business/coca-cola-pays-fortune-to-fallen-boss/FNXOZDNSDAEUP44G3N76PIWOMQ/

Ock, J., & Oswald, F. L. (2018). The utility of personnel selection decisions: Comparing compensatory and multiple-hurdle selection models. *Journal of Personnel Psychology, 17*, 172–182.

Oosterbeek, H., Sloof, R., & Van De Kuilen, G. (2004). Cultural differences in ultimatum game experiments: Evidence from a meta-analysis. *Experimental Economics, 7*, 171–188.

Ostrom, E. (1990). *Governing the commons: The evolution of institutions for collective action*. Cambridge University Press.

Pachur, T. (2022). Strategy selection in decisions from givens: Deciding at a glance? *Cognitive Psychology, 136*, Article 101483.

Pahl, G., & Beitz, W. (1996). *Engineering design: A systematic approach* (2nd ed.). Springer.

Pearl, J. (1984). *Heuristics: Intelligent search strategies for computer problem solving*. Addison-Wesley Longman.

Pólya, G. (1945). *How to solve it*. Princeton University Press.

Popomaronis, T. (2020, October 22). *Jeff Bezos's 3-question rule for hiring new Amazon employees—and how to answer them right*. CNBC. https://www.cnbc.com/2020/10/20/jeff-bezos-3-question-rule-for-hiring-new-amazon-employees.html

Popomaronis, T. (2021, January 27). *Elon Musk asks this question at every interview to spot a liar—why science says it actually works*. CNBC. https://www.cnbc.com/2021/01/26/elon-musk-favorite-job-interview-question-to-ask-to-spot-a-liar-science-says-it-actually-works.html

Posner, R. A. (2009). *A failure of capitalism*. Harvard University Press.

Proudfoot, D., & Lind, E. A. (2015). Fairness heuristic theory, the uncertainty management model, and fairness at work. In R. Cropanzano & M. L. Ambrose (Eds.), *Oxford handbook of justice in the workplace* (pp. 371–385). Oxford University Press.

Quigley, T. J., Hambrick, D. C., Misangyi, V. F., & Rizzi, G. A. (2019). CEO selection as risk taking: A new vantage on the debate about the consequences of insiders versus outsiders. *Strategic Management Journal, 40*, 1453–1470.

Rackham, N. (2007). The behavior of successful negotiators. In R. J. Lewicki, D. M. Saunders, & B. Barry (Eds.), *Negotiation: Readings, exercises, and cases* (5th ed., pp. 171–182). McGraw-Hill Irwin.

Rajshekhar, R. (2021, October 18). *What makes Apple design so good*. Medium. https://medium.com/macoclock/what-makes-apple-design-so-good-d430ef97c6d2

Randolph, M. (2019). *That will never work: The birth of Netflix and the amazing life of an idea*. Little, Brown.

Rapoport, A., & Chammah, A. M. (1965). *Prisoner's dilemma*. University of Michigan Press.

Rieskamp, J., & Otto, P. E. (2006). SSL: A theory of how people learn to select strategies. *Journal of Experimental Psychology: General, 135*, 207–236.

Roberts, M., Driggs, D., Thorpe, M., Gilbey, J., Yeung, M., Ursprung, S., Aviles-Rivero, A. I., Etmann, C., McCague, C., Beer, L., Weir-McCall, J. R., Teng, Z., Gkrania-Klotsas, E., AIX-COVNET, Rudd, J. H. F., Sala, E., & Schönlieb, C. B. (2021). Common pitfalls and recommendations for using machine learning to detect and prognosticate for COVID-19 using chest radiographs and CT scans. *Nature Machine Intelligence, 3*, 199–217.

Ross, C. (2022, February 28). *AI gone astray: How subtle shifts in patient data send popular algorithms reeling, undermining patient safety*. STAT. https://www.statnews.com/2022/02/28/sepsis-hospital-algorithms-data-shift/

Rudin, C. (2019). Stop explaining black box machine learning models for high stakes decisions and use interpretable models instead. *Nature Machine Intelligence, 1*, 206–215.

Sackett, P. R., & Lievens, F. (2008). Personnel selection. *Annual Review of Psychology, 59*, 419–450.

Salas, E., Shuffler, M. L., Thayer, A. L., Bedwell, W. L., & Lazzara, E. H. (2015). Understanding and improving teamwork in organizations: A scientifically based practical guide. *Human Resource Management, 54*, 599–622.

Salinas, S., & Meredith, S. (2018, October 24). *Tim Cook: Personal data collection is being 'weaponized against us with military efficiency.'* CNBC. https://www.cnbc.com /2018/10/24/apples-tim-cook-warns-silicon-valley-it-would-be-destructive-to-block -strong-privacy-laws.html

Savage, L. J. (1954). *The foundations of statistics.* John Wiley & Sons.

Schein, E. H. (1985). *Organizational culture and leadership.* Jossey-Bass Publishers.

Schmidt, F. L., & Hunter, J. E. (1998). The validity and utility of selection methods in personnel psychology: Practical and theoretical implications of 85 years of research findings. *Psychological Bulletin, 124*, 262–274.

Schumpeter, J. (1911). *Theorie der wirtschaftlichen Entwicklung.* Duncker and Humbolt.

Schumpeter, J. (1942). *Capitalism, socialism and democracy.* Harper.

Schweinsberg, M., Ku, G., Wang, C. S., & Pillutla, M. M. (2012). Starting high and ending with nothing: The role of anchors and power in negotiations. *Journal of Experimental Social Psychology, 48*, 226–231.

Science History Institute. (2020, September 21). *Robert W. Gore.* https://www .sciencehistory.org/historical-profile/robert-w-gore

Seifert, C. M., Gonzalez, R., Yilmaz, S., & Daly, S. (2016). Boosting creativity in idea generation using design heuristics. In Product Development and Management Association (Ed.), *Design and design thinking* (pp. 71–86). Wiley.

Selten, R. (1978). The chain store paradox. *Theory and Decision, 9*, 27–59.

Serwe, S., & Frings, C. (2006). Who will win Wimbledon? The recognition heuristic in predicting sports events. *Journal of Behavioral Decision Making, 19*, 321–332.

Shah, A. K., & Oppenheimer, D. M. (2008). Heuristics made easy: An effort-reduction framework. *Psychological Bulletin, 134*, 207–222.

Shankar, V., & Carpenter, G. (2012). Late-mover strategies. In V. Shankar & G. Carpenter (Eds.), *Handbook of marketing strategy* (pp. 362–375). Edward Elgar.

Sharapov, D., & Ross, J. M. (2023). Whom should a leader imitate? Using rivalry-based imitation to manage strategic risk in changing environments. *Strategic Management Journal, 44*, 311–342.

Sherden, W. A. (1998). *The fortune sellers: The big business of buying and selling predictions.* John Wiley & Sons.

Siedel, G. J. (2014). *Negotiating for success: Essential strategies and skills*. Van Rye Publishing.

Simon, H. A. (1947). *Administrative behavior: A study of decision-making processes in administrative organizations*. Macmillan.

Simon, H. A. (1955). A behavioral model of rational choice. *Quarterly Journal of Economics, 69*, 99–118.

Simon, H. A. (1971). Designing organizations for an information-rich world. In M. Greenberger (Ed.), *Computers, communications, and the public interest* (pp. 37–72). Johns Hopkins University Press.

Simon, H. A. (1988). Nobel laureate Simon "looks back": A low-frequency mode. *Public Administration Quarterly, 12*, 275–300.

Softbank. (2000). *2000 Annual report*. https://group.softbank/system/files/pdf/ir /financials/annual_reports/annual-report_fy2000_01_en.pdf

Stewart, J. B. (2012, May 18). In the undoing of a C.E.O., a puzzle. *New York Times*. https://www.nytimes.com/2012/05/19/business/the-undoing-of-scott-thompson-at -yahoo-common-sense.html

Stigler, S. M. (1980). Stigler's law of eponymy. In F. Gieryn (Ed.), *Science and social structure: A festschrift for Robert K. Merton* (pp. 147–158). New York Academy of Sciences.

Stone, P. (2011). *The luck of the draw: The role of lotteries in decision making*. Oxford University Press.

Studdert, D. M., Mello, M. M., Sage, W. M., Desroches, C. M., Peugh, J., Zapert, K., & Brennan, T. A. (2005). Defensive medicine: Among high-risk specialist physicians in a volatile malpractice environment. *JAMA, 293*, 2609–2617.

Sull, D., & Eisenhardt, K. M. (2015). *Simple rules: How to thrive in a complex world*. Houghton Mifflin Harcourt.

Swaab, R. I., Maddux, W. W., & Sinaceur, M. (2011). Early words that work: When and how virtual linguistic mimicry facilitates negotiation outcomes. *Journal of Experimental Social Psychology, 47*, 616–621.

Taleb, N. N., & Blyth, M. (2011). The black swan of Cairo: How suppressing volatility makes the world less predictable and more dangerous. *Foreign Affairs, 90*, 33–39.

Tan, J. H., Luan, S., & Katsikopoulos, K. (2017). A signal-detection approach to modeling forgiveness decisions. *Evolution and Human Behavior, 38*, 27–38.

Tavris, C., & Aronson, E. (2007). *Mistakes were made (but not by me): Why we justify foolish beliefs, bad decisions, and hurtful acts*. Harcourt.

Tetlock, P. E. (2003). Thinking the unthinkable: Sacred values and taboo cognitions. *Trends in Cognitive Sciences, 7*, 320–324.

Tey, K. S., Schaerer, M., Madan, N., & Swaab, R. I. (2021). The impact of concession patterns on negotiations: When and why decreasing concessions lead to a distributive disadvantage. *Organizational Behavior and Human Decision Processes, 165*, 153–166.

Thaler, R. H., & Sunstein, C. R. (2008). *Nudge: Improving decisions about health, wealth, and happiness.* Yale University Press.

Thomadsen, R. (2007). Product positioning and competition: The role of location in the fast food industry. *Marketing Science, 26*, 792–804.

Thuderoz, C. (2017). Why do we respond to a concession with another concession? Reciprocity and compromise. *Negotiation Journal, 33*, 71–83.

Todd, P. M., & Miller, G. F. (1999). From pride and prejudice to persuasion: Satisficing in mate search. In G. Gigerenzer, P. M. Todd, & the ABC Research Group, *Simple heuristics that make us smart* (pp. 287–308). Oxford University Press.

Tomasello, M. (2019). *Becoming human: A theory of ontogeny.* Harvard University Press.

Turek, M. (n.d.). *Explainable artificial intelligence (XAI).* Defense Advanced Research Projects Agency (DARPA). https://www.darpa.mil/program/explainable-artificial-intelligence

Tversky, A., & Kahneman, D. (1974). Judgment under uncertainty: Heuristics and biases. *Science, 185*, 1124–1131.

United Nations. (n.d.). *COP26: Together for our planet.* United Nations Climate Action. https://www.un.org/en/climatechange/cop26

US Department of Justice. (2022, May 12). *Justice Department and EEOC warn against disability discrimination.* https://www.justice.gov/opa/pr/justice-department-and-eeoc-warn-against-disability-discrimination

US Equal Employment Opportunity Commission. (2022, May 12). *The Americans with Disabilities Act and the use of software, algorithms, and artificial intelligence to assess job applicants and employees.* https://www.eeoc.gov/laws/guidance/americans-disabilities-act-and-use-software-algorithms-and-artificial-intelligence

van den Bos, K., & Lind, E. A. (2002). Uncertainty management by means of fairness judgments. In M. P. Zanna (Ed.), *Advances in experimental social psychology* (Vol. 34, pp. 1–60). Academic Press.

van Dyck, C., Frese, M., Baer, M., & Sonnentag, S. (2005). Organizational error management culture and its impact on performance: A two-study replication. *Journal of Applied Psychology, 90*, 1228–1240.

van Vugt, M., Hogan, R., & Kaiser, R. B. (2008). Leadership, followership, and evolution: Some lessons from the past. *American Psychologist, 63*, 182–196.

van Vugt, M., Johnson, D. D. P., Kaiser, R. B., & O'Gorman, R. (2008). Evolution and the social psychology of leadership: The mismatch hypothesis. In C. Hoyt, G. R. Giethals, & D. R. Forsyth (Eds.), *Leadership at the crossroads: Leadership and psychology* (Vol. 1, pp. 267–282). Praeger.

Vitsoe. (n.d.). *The power of good design.* https://www.vitsoe.com/us/about/good-design

Vroom, V. H., & Jago, A. G. (1988). *The new leadership: Managing participation in organizations.* Prentice Hall.

Vroom, V. H., & Jago, A. G. (2007). The role of the situation in leadership. *American Psychologist, 62*, 17–24.

Wade, B. (1988, November 20). Practical traveler; mileage points can fly away. *New York Times.* https://www.nytimes.com/1988/11/20/travel/practical-traveler-mileage-points-can-fly-away.html

Walther, J. B., & Bunz, U. (2005). The rules of virtual groups: Trust, liking, and performance in computer-mediated communication. *Journal of Communication, 55*, 828–846.

Walumbwa, F. O., Maidique, M. A., & Atamanik, C. (2014). Decision-making in a crisis: What every leader needs to know. *Organizational Dynamics, 44*, 284–293.

Wegwarth, O., Gaissmaier, W., & Gigerenzer, G. (2009). Smart strategies for doctors and doctors-in-training: Heuristics in medicine. *Medical Education, 43*, 721–728.

West, D. C., Acar, O. A., & Caruana, A. (2020). Choosing among alternative new product development projects: The role of heuristics. *Psychology & Marketing, 37*, 1511–1524.

White, A. (2019, January 3). Our top data and analytics predicts for 2019. *Gartner.* https://blogs.gartner.com/andrew_white/2019/01/03/our-top-data-and-analytics-predicts-for-2019

Whitfield, J. (2008). Collaboration: Group theory. *Nature News, 455*, 720–723.

Wikipedia. (n.d.). *Wikipedia: Consensus.* https://en.wikipedia.org/wiki/Wikipedia:Consensus

Wong, A., Otles, E., Donnelly, J. P., Krumm, A., McCullough, J., DeTroyer-Cooley, O., Pestrue, J., Philips, M., Konye, J., Penoza, C., Ghous, M., & Singh, K. (2021). External validation of a widely implemented proprietary sepsis prediction model in hospitalized patients. *JAMA Internal Medicine, 181*, 1065–1070.

Woolley, A. W., Chabris, C. F., Pentland, A., Hashmi, N., & Malone, T. W. (2010). Evidence for a collective intelligence factor in the performance of human groups. *Science, 330*, 686–688.

Wübben, M., & von Wangenheim, F. (2008). Instant customer base analysis: Managerial heuristics often "get it right." *Journal of Marketing, 72,* 82–93.

Wynants, L., Calster, B. V., Collins, G. S., Riley, R. D., Heinze, G., Schuit, E., Bonten, M. M. J., Dahly, D. L., Damen, J. A. A., Debray, T. P. A., de Jong, V. M. T., De Vos, M., Dhirman, P., Haller, M. C., O'Harhay, M., Henckarts, L., Heus, P., Kammer, M., Kreusberger, N., . . . van Smeden, M. (2020). Prediction models for diagnosis and prognosis of Covid-19: Systematic review and critical appraisal. *British Medical Journal, 369,* Article m1328.

Xinhua News Agency. (2022, November 16). Bad loan ratio of China's commercial banks declines in first 3 quarters. *China Daily.* https://www.chinadaily.com.cn/a/202211/16/WS6374522ea31049175432a09d.html

Yang, J., & Tilley, A. (2022, December 3). Apple makes plans to move production out of China. *Wall Street Journal.* https://www.wsj.com/articles/apple-china-factory-protests-foxconn-manufacturing-production-supply-chain-11670023099

Yanofsky, N. S. (2013). *The outer limits of reason: What science, mathematics, and logic cannot tell us.* MIT Press.

Yilmaz, S., Seifert, C. M., & Gonzalez, R. (2010). Cognitive heuristics in design: Instructional strategies to increase creativity in idea generation. *Artificial Intelligence for Engineering Design, Analysis and Manufacturing, 24,* 335–355.

Zetlin, M. (n.d.). Blockbuster could have bought Netflix for $50 million, but the CEO thought it was a joke. *Inc.* https://www.inc.com/minda-zetlin/netflix-blockbuster-meeting-marc-randolph-reed-hastings-john-antioco.html

Index

1/N heuristic, 6–7, 19, 23, 35, 44–45,
    132, 162, 179
    1/N minus delta, 116
    definition, 211
    in hiring, 66
    in negotiation, 118
    in team communication, 122
    in ultimatum game, 116–117
15 percent rule, 92–93
30/4 rule, 93–94
3M (company), 92–95, 175
6 percent rule, 93–94

AB (adaptive boosting), 187
ABC research group, 133–135
Adaptive toolbox, 11–12, 18, 31, 33–52,
    204, 208. *See also* Ecological
    rationality; Heuristics
    classes of heuristics in the, 34–49
    culture and, 164
    definition, 211
    in hiring, 66, 69
    in leadership, 137–140, 148
    in negotiation, 119
    in project management, 141–143
    refining the, 207–210
    of strategy heuristics, 82–88
    teaching the, 198–201, 210
AI algorithms. *See* Machine learning
Algorithmic models of heuristics, 15–16.
    See also Heuristics

AlphaGo, 178
Amazon, 47, 74, 96, 100–101. *See also*
    Bezos, Jeff
Ambiguity, 17–19, 129. *See also* VUCA
    definition, 211
Amelio, Bill, 139
Apple, 39, 47, 78. *See also* Jobs, Steve
    iPhone, 77–78
    iPod, 97–98
Argument dilution, 109
Arison, Micky, 138
Aristotle, 143
Artificial intelligence (AI), 177–193,
    200–201. *See also* Machine
    learning
    AI winters, 178
    black-box algorithms, 191–193
    ecological rationality of,
        179–180
    explainable AI (XAI), 29
    in hiring, 65–66
    origins, 15
Artinger, Florian, 175
Aspiration-based heuristics, 35, 45–47.
    See also Satisficing
Aspiration level, 6, 35. *See also* Satisficing
    definition, 211
Attention economy, 146
Attention overload, 146
Avalanche risk, 43
Axelrod, Robert, 129

Backward induction, 8
Baiting heuristic, 86–87
Bank of England, 203
Barbie doll, 101
Baseball. *See* Gaze heuristic
Basel Accord, 169
Bayes's rule, 101, 167
   Bayesian decision theory, 18
   Bayesian linear regression, 42
   Bayesian models, 52
   Bayesian updating, 33, 110
   naive Bayes, 187
Bell, Graham, 92
Bezos, Jeff, 91, 100, 197. *See also* Amazon
   fast-and-frugal trees for hiring, 58–60
   two-pizza rule, 123
BG/NBD (beta geometric/negative
   binomial distribution) model,
   180–181
Bias, 11, 13, 48–49. *See also* Bias-variance
   dilemma; Heuristics-and-biases
   program; Naïve diversification bias
   bias bias, 106
   bias-free, 66
   debiasing hiring decisions, 67–69
   in negotiations, 104
Bias-variance dilemma, 51–52, 67
   definition, 211
Biden, Joe, 195
Bingham, Christopher, 87–88, 203, 206,
   208
Blackberry (company), 77–78
Black-box algorithms, 152, 190–193.
   See also Artificial intelligence;
   Machine learning
Blair, Tony, 195
Blockbuster (company), 90–92
Braun (company), 97–98
Brin, Sergey, 77, 197. *See also* Google
Budgeting, 162
Buffett, Warren, 39
Building blocks, 40–41, 58, 142. *See also*
   Heuristics

   definition, 211
Burger King, 86
Bush, George W., 195
Business school education, 3–4, 6, 8–9,
   18, 24–25, 119, 164, 195–210

Cake rule, 134
Cambiasso, Estéban, 158
Carlton, Richard, 95
Carpenter, Gregory, 79
CART (Classification and regression
   tree), 187
Casali, Erin, 123
Cauchy, Augustin-Louis, 101
Chain store paradox, 7–8, 10
Chamberlain, Neville, 195
ChatGPT. *See* Generative AI
Cisco, 83
Classification, 39, 43, 60, 186–188
Clinton, Bill, 195
Coca-Cola, 55, 80–82
Common knowledge, 113, 128
Cook, Tim, 193, 198
Coombs, Clyde, 124
Corrigan, Bernard, 44
COVID–19, 3, 41, 190–191, 198
Creative destruction, 91
Crew resource management training, 174
Cross-validation, 189. *See also* Out-of-
   sample prediction; Prediction
   definition, 211
Crowdsourcing, 100, 129
Culture, 111, 130, 133–135, 147–148,
   156, 161. *See also* Decision-making
   culture; Error culture
CYA culture, 164–166. *See also* Decision-
   making culture; Defensive decision
   making

Daoism, 139–140
Darwin, Charles, 173
Data shift, 191
Dawes, Robyn, 44

Decision by ordeal, 161
Decision-making culture, 12, 148,
    161–176
  CYA culture, 164–166
  definition, 162
  positive error culture, 171, 173–174
  positive heuristic culture, 171–173
  positive VUCA culture, 170–172
  rationalization culture, 163–164
  smart decision-making culture,
    170–176
  Turkey illusion culture, 166–168
  VUCA-denial culture, 168–170
Defensive decision making, 90, 148,
    154–156, 164–166, 176. *See also* CYA
    culture; Decision-making culture
  definition, 211
Deliberation, 12, 23–25, 151–154, 163.
    See also Intuition
Delta-inference, 35, 41–42, 51
  definition, 211
  in hiring decisions, 61–63, 184–185
Dilemmas of trust and honesty,
    112–113. *See also* Negotiation.
Dominant-cue condition, 49–51, 57,
    208. *See also* Ecological rationality
  definition, 211
  in hiring decisions, 62–64, 67
Drawing lots, 131–132
Driggs, Dereck, 190
Drosophila (fruit flies), 18
Drucker, Peter, 137, 161
Dual-system theories, 22, 151, 159.
    See also System 1 versus System 2
Dubey, Abeer, 121
Duncker, Karl, 15

Early wins heuristic, 122
Ecological rationality, 12, 16, 29, 34,
    37, 49–52, 57, 70–71, 88, 178,
    198–200, 206. *See also* Adaptive
    toolbox; Smart heuristics
  definition, 212

experience and, 62–64
  of leadership heuristics, 137, 140–141,
    143
  in negotiations, 106, 109, 117–118
  selecting heuristics, 207–210
  of tit-for-tat, 115–116
  trial-and-error, 96
Edison, Thomas, 95
Effort-accuracy trade-off, 22, 25, 44, 62
  definition, 212
Einstein, Albert, 15, 151–152, 159
Eisenhardt, Kathleen, 39, 87, 203, 206,
    208
Epic Systems (company), 190–191
Equal Employment Opportunity
    Commission, 65
Equality heuristics, 42–45, 49–51,
    126, 144. *See also* 1/*N*; Tallying;
    Unit-weighting;
  definition, 212
  in negotiation, 117–118
Error culture
  definition, 212
  in education, 174
  error management, 175
  error prevention, 175
  negative error culture, 25, 90, 156,
    159, 171, 173
  positive error culture, 94–96, 133,
    171, 173–176
Expected utility maximization, 3–5,
    10–11, 16–18, 21, 30–31, 33, 146,
    163, 168, 198–199
Experience, 23–25, 34, 38, 79, 103, 146,
    151–152, 153–157, 159, 180, 192,
    196, 206, 209. *See also* Fluency
    heuristic; Intuition
  learning from, 88, 95, 206–208,
    210
  in loan business, 185–186
  in personnel decisions, 62–64, 70
  in underwriting, 203–205
Experimental economics, 113

Facebook, 87, 89, 96, 197

Failure-to-success heuristic, 94

Fairchild, David, 137

Fairness heuristic theory, 145

False negatives, 60, 186–188
  false-negative rate, definition,
    212

False positives, 60–61, 186–188, 190
  false-positive rate, definition,
    212

Fast-and-frugal heuristics program,
    11–12, 16, 21, 34, 201. *See also*
    Heuristics; Smart heuristics

Fast-and-frugal trees, 35, 39–41, 84–85,
    203, 208–209
  definition, 212
  for hiring, 58–60
  in loan decisions, 186–189
  in performance management
    decisions, 69–70, 208–209
  in teams, 127–128

Fear index, 167–168

First listen then speak, 48, 122, 138,
    141, 148, 204

Fitting, 5, 22, 189. *See also* Overfitting;
    Prediction
  definition, 212

Fluency heuristic, 23–25, 35, 37, 96,
    156–158
  definition, 212

Flyvbjerg, Bent, 142–143, 206–207

Follower's dilemma, 145

Forced distribution performance
    management. *See* Stack ranking

Fourier, Joseph, 101

Franklin, Benjamin, 3–4

Friedman, Milton, 5, 169

Galton, Francis, 48

Game theory, 7, 10, 104, 113

Gandhi, Mahatma, 148

Gaze heuristic, 38–39, 202
  definition, 212

Generative AI, 182–184

Goizueta, Roberto, 55

Gombart, Werner, 91

Google, 47, 89, 91, 178, 197–198
  20 percent time, 93, 162
  AdWords, 87, 101
  business model, 87
  Google Flu Trends, 26–29
  imitation, 77
  Project Aristotle, 121–122
  Project Unbias, 67–69

Gore-Tex, 96

Grammar of norms, 132

Gross-margin heuristic, 84

Grove, Andy, 140–141

Gupta, Rajat, 196

Gut feelings. *See* Intuition

Hägeli, Pascal, 43

Hamilton, Alexander, 75

Harnack principle, 133, 138

Harvard Business School, 195–197

Hastings, Reed, 90

Heuristics, 6–9, 11–12. *See also*
    Adaptive toolbox; Ecological
    rationality; Fast-and-frugal
    heuristics program; Heuristics-and-
    biases program; Smart heuristics
  advantages of, 21–31
  classes of, 34
  common misconceptions, 30–31
  definition, 4, 212
  ecological rationality of, 49–52
  evolved heuristics, 201–202
  learning of, 201–210
  origin, 15, 89
  portfolio of, 203

Heuristics-and-biases program, 16
  in business school education, 199
  in negotiations, 104

Hewlett, Bill, 93

Hiatus heuristic, 25–29, 180–182
  definition, 212

Hiring decisions, 33–34, 55–69, 138,
    184–185, 203, 208
  costs of bad hiring, 55–56
  heuristics for, 56–65
  multiple-hurdle selection, 60–61
  number of interviewers, 66–68
Hoover, Herbert, 137
Hostage negotiation, 103, 117
Huawei, 131

IBM, 177–179
Illusion of certainty, 3–4, 167
  definition, 212
Illusion of complexity, 189
Imitate-the-majority, 63, 79–80, 86,
    201–202. *See also* Imitation
  definition, 212
Imitate-the-successful, 79, 97–98, 110,
    122, 201. *See also* Imitation
  definition, 213
Imitation, 35, 47–48, 63, 74, 101.
    See also Imitate-the-majority;
    Imitate-the-successful; Mirror
    heuristic; Time machine
    heuristic
  economic development, 75–77
  in economic games, 115–116
  innovative imitation, 77–78
  learning and, 47, 201–204
  in negotiation, 110–111
  pirating, 76–77
  strategy, 73–80
Insurance underwriting, 174,
    203–205
Intel, 81–82, 84, 140–141
Intelligence, 11, 15, 21, 34, 135, 152,
    154, 179, 192. *See also* Artificial
    intelligence
  emotional intelligence, 55
Intractability, 10, 17–18, 20, 30–31, 179,
    199–200. *See also* Large Worlds;
    VUCA
  definition, 213

Intuition, 8–9, 11, 21, 24–25, 30, 67,
    71, 147, 151–159, 162, 164, 172,
    200–201
  blocking of, 158
  definition, 152, 213
  fear of admitting, 153–154
Ivester, Douglas, 55

Jackson, Samuel, 103
Jago, Arthur, 136
Jennings, Ken, 177
Jeopardy!, 177–179
Jesuits. *See* Society of Jesus
Jobs, Steve, 78, 132, 135, 197. *See also*
    Apple
Johnson, Boris, 195

Kahneman, Daniel, 19. *See also*
    Heuristics-and-biases program
Katsikopoulos, Konstantinos, 188, 203
Keppler, Johannes, 47
Keys to the White House model, 43
KFC, 80–81, 201–202
Klein, Gary, 156–157
Knight, Frank, 4, 9, 21, 170
KNN (k-nearest neighbor), 187
Kohl, Helmut, 195

Lampel, Joseph, 197
Laplace, Pierre-Simon, 101
Large worlds, 16–22, 26, 30–31, 33,
    104, 106, 109, 162, 169–171, 179,
    182, 184, 188, 192. *See also* Small
    worlds; Uncertainty
  definition, 213
LASSO regression, 182–185
Late-mover heuristic, 78–79. *See also*
    Imitation; Used apple policy
Law of eponymy, 101
Layoff decision, 40–41, 69–70
Leader's adaptive toolbox, 133–148.
    See also Leadership
  definition, 213

Leadership, 133–148
  benefits of leadership heuristics,
    145–148
  as decision making, 137–138
  choosing leaders, 143–145
  constructive leadership heuristics, 139
  contingency model, 136
  destructive leadership heuristics, 139
  ecological rationality, 140–141
  evolutionary leadership, 135,
    144–145
  leadership styles, 136–137
  leadership theories, 135–137
  proverbs, 139–140
Learning, 69, 76–77, 89, 133–134, 143,
    164, 166, 175. *See also* Imitation;
    Machine learning
  of decision making, 210
  from failure, 95
  of heuristics, 47, 201–208
  of how to select heuristics, 208–210
  opportunities for, 62–63, 185
Lego (company), 142
  LEGO ideas, 100
Lehman Brothers, 168
Lehmann, Jens, 158
Less-is-more effect, 7, 11, 22, 25, 62, 67,
    109, 143, 171, 180, 197
  definition, 213
Levitt, Theodore, 77, 79, 83
Lichtman, Allan, 43
Logistic regression, 62–63, 70, 180–181,
    184–185, 187
Lowell, Francis, 76
Lucas, Robert, 20, 167
Luce, Duncan, 19

Ma, Li, 139
Machine learning, 15, 25, 28–29, 133,
    178–189, 203
Maddux, William, 110
Madoff, Bernie, 139
Maidique, Mitch, 138–139

Maistry, Gavin, 203
Management education. *See* Business
    school education
Mandela, Nelson, 148
Manimala, Mathew, 89
Mannes, Albert, 124
March, James, 137
Markowitz, Harry, 6–7, 9–11, 19, 44
Marx, Karl, 91
Masterbuilder heuristics, 142–143
Mathematics education, 174, 200
Mattel, 101
Maximizing, 4, 6, 16, 93, 162. *See also*
    Expected utility maximization
Max Planck Institute for Human
    Development, 203
Max Planck Institute for Psychological
    Research, 133
Max Planck Society, 133, 138
May, Theresa, 195
MBA. *See* Business school education
McCammon, Ian, 43
McDonald's, 80, 86
McKnight Principles, 95
Mean-variance portfolio, 6–7, 23, 44–45
Meeting-in-the-middle heuristic, 118
Megaprojects, 141–143
Merkel, Angela, 195
Merton, Robert, C., 10
Merton, Robert, K., 101
Microsoft, 71
Middle-born-child effect, 45
Mintzberg, Henry, 146, 195–198
Miramax, 84–85
Mirror heuristic, 111
Moore, Gordon, 140–141
More is always better belief, 3, 26, 66,
    164, 170–171, 198. *See also* Less-is-
    more effect
Multiplier heuristic, 181–183, 189
Musk, Elon, 135, 197
  hiring rule, 33–34, 37, 49, 56–58,
    203

Naïve diversification bias, 7
National Football League, 41–42, 125
Naturalistic decision making, 156–157
Nearest-neighbor heuristic, 18, 201
Negotiation, 103–119
    asking questions, 108
    behavior labelling, 108
    counterproposal, 107
    distributive (win-lose) negotiation,
        110
    first offer, 118
    integrative (win-win) negotiation, 118
    irritator words, 107
    preparation, 107
    reciprocal concession making, 112
    strategies, 109
    verbal attack, 107
Negotiator, the (movie), 103, 117
Netflix, 89–92
Neural network, 179, 182, 187–188
Newell, Allen, 15, 182
Nixon, Richard, 195–196
No-trade theorem, 21

Obama, Barack, 195–196
Obvious clues method, 43
One-clever-cue heuristics, 33, 35–39, 50,
        52, 84, 127, 139, 180, 203–204
    definition, 213
    in hiring, 56–57
    for innovation, 92–94
One-reason decision making, 35, 37–42,
        49–50, 63–64, 86, 126, 208, 210.
        See also Delta-inference; Dominant-
        cue condition; Fast-and-frugal
        trees; One-clever-cue heuristics;
        Take-the-best
    definition, 213
Optimization, 5, 9, 11–12, 20, 30, 34,
        43, 84, 86, 106, 114, 162–164,
        168–170, 199. See also Expected
        utility maximization; Small worlds
    definition, 213

Organizational culture. See also
        Decision-making culture;
        Defensive decision making; Error
        culture
    leadership and, 147–148
Ostrom, Elinor, 130–132
Out-of-population prediction, 189–191.
        See also Prediction
    definition, 213
Out-of-sample prediction, 5, 189–190.
        See also Cross-validation; Prediction
    definition, 213
Overfitting, 7, 26–28, 30, 44, 136, 189.
        See also Prediction

Page, Larry, 77, 197. See also Google
Pareto/NBD (negative binomial
        distribution) model, 180–181
Paulson, Henry, 167
PayPal, 89, 126
Pearl, Judea, 200
Performance management, 69–71.
        See also Stack ranking
Phronesis, 143
Pichai, Sundar, 198
Poisson, Siméon, 101
Pólya, George, 15, 200
Ponzi scheme, 139
Positive VUCA culture, 170–172
Post-It Note, 92–94, 175
Powell, David, 94
Prediction, 7, 31, 41, 43–44, 52, 92, 114,
        116, 167, 182–184, 193
    accuracy of, 22, 26–29, 36, 41, 99,
        180–181
    error, 52, 182
    out-of-population, 189–191,
        213
    out-of-sample, 5, 189–190, 213
Priority heuristics, 87
Prisoner's dilemma, 114–116, 129
Procedural heuristics, 87–88
Product design heuristics, 96–99

Profit, 5–6, 9, 24, 45, 74–75, 78, 84, 86, 94, 100, 123, 173, 203. *See also* Risk; Uncertainty
    under risk and uncertainty, 21, 170–171
Project management, 141–143
Psychological AI, 15, 177–184. *See also* Artificial intelligence

Rackham, Neil, 106, 109
Raiffa, Howard, 19
Rams, Dieter, 97
Random forest, 25, 28–29, 31, 179–188
Rank and yank. *See* Stack ranking
Rapoport, Amnon, 113
Rationalization culture, 154, 163–164, 166, 169
Receiver-operating curve, 186–187
Recency heuristic, 26–29, 162
    definition, 213
Reciprocal concession heuristic, 112
Reciprocity, 122. *See also* Tit-for-tat
    in negotiation, 111–113
    positive reciprocity, 111
    reciprocity heuristic, 111
Recognition heuristic, 34–37, 165.
    definition, 213–214
Recognition-primed decision making, 157
Redundancy, 57
Ridge regression, 182–183
Risk, 9, 17, 19, 25, 43, 61, 65. *See also* Risk versus uncertainty; Small worlds
    decision-making culture and, 155–156, 159, 163, 167
    definition, 214
    innovation and, 77, 90–91, 94
    in leadership, 133, 145
    of loans, 186–188
    in negotiation, 110
    in underwriting, 203–204

Risk versus uncertainty, 3–4, 6, 9–11, 17, 19–21, 25, 30–31, 91, 168–169, 199, 210
    and innovation, 21, 93, 170
    and profit, 21, 170
Ross, Jan-Michael, 82
Rule of five, 123

Sacred values, 126–127
Salas, Eduardo, 122
Samwer brothers, 74
Satisficing, 4–6, 15–16, 35, 45–47, 83, 122, 158, 173, 210
    with and without aspiration-level adaptation, 45–46, 84–86
    definition, 214
    in negotiations, 110, 112
Savage, Leonard, 16–18
Schumpeter, Joseph, 91
ScotchGuard, 94–95
Search rule, 40, 59, 200
Second-mover heuristic, 81
Secretary problem, 46–47
Selection heuristics, 87, 206
Selten, Reinhard, 7–11, 20, 106
Sharapov, Dmitry, 82
Sherman, Patsy, 94
Simon, Herbert, 4–6, 9–12, 15–16, 20, 45, 110, 137, 146, 179, 182, 192
Single-peaked preference function, 124–125
Skilling, Jeffrey, 139, 196
Slater, Samuel, 75–76
Small worlds, 16–22, 30–31, 46, 91, 93, 118, 193, 198. *See also* Risk; Stable-world principle
    decision-making cultures and, 164–173
    definition, 214
    in games, 113–116
    in negotiation research, 104–106
Smart heuristics, 3, 11, 18, 71, 165, 172.
    See also Ecological rationality
    artificial intelligence and, 192–193

business school education and, 198–200
definition, 12, 214
in hiring, 62, 66–67, 69
in innovation, 89, 101
in leadership, 138, 146–148
learning, 203–208
in performance management, 69–70
in social dilemmas, 129–132
in strategy, 88
Smith, Vernon, 113
Social heuristics, 33, 35, 47–49, 202, 210. *See also* Imitation; Word-of-mouth; Wisdom-of-crowds
definition, 214
for hiring, 63–65
Society of Jesus, 171–172
Socrates, 143
SoftBank, 73–74
Son, Masayoshi, 73–74, 201
Sopranos, the (TV show), 127
Spacey, Kevin, 117
Speed-accuracy trade-off, 22–25, 62, 157
definition, 214
faster-is-better principle, 23
Spolsky, Joel, 123
Stable-world principle, 179–180, 192. *See also* Risk; Small worlds
definition, 214
Stack ranking, 70–71
Stata, Ray, 139
Stigler, Stephen, 101
Stone, Peter, 131
Stopping rule, 40, 46, 59, 61, 67, 200
Strategy, 18, 39, 46, 73–88, 129, 154, 161, 172, 181, 199, 201–202. *See also* Heuristics
acquisition strategy, 83
hiring strategy, 56–60
location strategy, 86

market expansion strategy, 86–87
negotiation strategies, 108–117
pricing strategy, 84–85
production strategy, 83–84
Sull, Donald, 39
Suntory, 80–82
Sun Tzu, 88
SVM (support vector machine), 184–185, 187–188
System 1 versus System 2, 23, 30, 69, 71, 151. *See also* Deliberation; Intuition

Table rule, 123
Take-the-best heuristic, 35, 39, 41, 50, 52
definition, 214
Take-the-first heuristic. *See* Fluency heuristic.
Tallying, 35, 43, 50, 83, 85, 97, 126, 203, 208. *See also* Equality heuristics
definition, 214
Target-reservation heuristic, 110. *See also* Negotiation
Tax rates, 162–163
Teams, 121–132
bad apples in a team, 126–128
heuristics for effective teamwork, 122–124
heuristics for virtual teams, 128–129
Temporal heuristics, 87–88, 206
Tetlock, Philip, 126–127
Thatcher, Margaret, 195
Thomadsen, Raphael, 86
Thompson, Scott, 126
Time machine heuristic, 73–75, 79, 201
Tit-for-tat, 114–116, 122, 129. *See also* Negotiation; Reciprocity
definition, 214
tit-for-two-tats, 115
Tokyo Summer Olympics, 141–142
Tragedy of the commons, 129–132

Transparency, 28–29, 200. *See also*
    Transparency-accuracy trade-off
  AI and, 191–193
  definition, 214
  discrimination and, 65–66
  in hiring decisions, 65–66, 69
  in leadership, 147
Transparency-accuracy trade-off, 22,
    28–29, 62
  definition, 215
Traveling salesperson problem, 18, 201.
    See also Intractability; Nearest-
    neighbor heuristic
Trial-and-error, 95
  ecological rationality of, 96
  heuristic, 95
Trump, Donald, 43, 195
Tsui, Anne, 139
Turkey illusion culture, 166–169
Turn-taking, 130–131
Tversky, Amos, 19. *See also* Heuristics-
    and-biases program
Two-pizza rule, 123

Ultimatum game, 116–117
Uncertainty, 3, 6, 9–11, 17, 63, 69, 80,
    88, 91, 130, 133, 170, 174, 177.
    See also Large worlds; Risk versus
    uncertainty; VUCA
  definition, 215
  ecological rationality and, 49–52
  entrepreneurship and, 171
  heuristics and, 4, 16–19, 22, 25, 28,
    30–31, 44, 46, 49
  innovation and, 20–21, 90–92,
    170–171
  intuition and, 151–152
  irreducible, 168–169, 171
  machine learning and, 184–185,
    192
  negotiation and, 104–109
  profits and, 20–21, 170–171
  radical uncertainty, 17

  selecting leaders and, 145
  stable-world principle and, 179
  teaching heuristics and, 198–200,
    202, 210
  VUCA and, 19–20
  VUCA-denial culture and, 168–169
Unit weighting, 35, 43–44, 126
  definition, 215
Ur-heuristic, 143
US presidents, 137, 195–196
  US presidential election, 43
Used apple policy, 79
Uzzi, Brian, 123

van Dyck, Cathy, 175
Variance. *See* Bias-variance dilemma;
    Mean-variance portfolio; VUCA
Venkatesh, Shankar, 79
Vision Fund, 73
von Clausewitz, Carl, 88
von Harnack, Adolf, 133
von Wangenheim, Florian, 180
Vroom, Victor, 136
VUCA, 4, 11, 75, 168–170, 189, 199,
    207, 210. *See also* Large worlds;
    Uncertainty
  definition, 19–20, 215
  positive VUCA culture, 170–172
VUCA-denial culture, 168–170

Watson (IBM supercomputer),
    177–179
Welch, Jack, 70, 135, 151
Wertheimer, Max, 15
Wheelan, Susan, 124
Wikipedia, 129
Wimbledon, 35–36
Win-stay, lose-shift, 115–116
Wintering rule, 130
Wisdom-of-crowds, 35, 48–49, 100
  definition, 215
  innovation, 99–100
  wisdom of select crowds, 124–125

Word-of-mouth, 33–35, 48, 63–65
    definition, 215
Wübben, Markus, 180

Xiaomi, 78

Yahoo!, 74, 126
Yamamotoyama (company), 172–173

Zuckerberg, Mark, 197